EOS

HERMES

SIBYLLA

ARCHIMEDES

TO-DAY AND TO-MORROW

Volume 9

EOS

By J. H. Jeans

HERMES

By T. W. Jones

SIBYLLA

By C. A. Mace

ARCHIMEDES

By L. L. Whyte

Routledge
Taylor & Francis Group
LONDON AND NEW YORK

Edition Synapse

First published in 1928, 1928, 1927, 1928

This edition first published in 2008 by
Routledge
2 Park Square, Milton Park, Abingdon, Oxon, OX14 4RN

Simultaneously published in the USA and Canada
by Routledge
605 Third Avenue, New York, NY 10017

Simultaneously published in Japan
by Edition Synapse, 2-8-5 Uchikanda, Chiyoda-ku, Tokyo
101-0047

Routledge is an imprint of the Taylor & Francis Group, an informa business

British Library Cataloguing in Publication Data
A catalogue record for this book is available from the British Library

Publisher's Note
The publisher has gone to great lengths to ensure the quality of this reprint but points out that some imper-fections in the original copies may be apparent.

Disclaimer
The Publishers have made every effort to trace copy-right holders and would welcome correspondence from those they have been unable to contact.

ISBN 13: 978-0-415-43596-3 (Set)
ISBN 13: 978-0-415-46304-1 (hbk) (vol 9)

Japan ISBN 13: 978-4-86166-060-3 (Set)
Japan ISBN 13: 978-4-86166-070-2 (hbk) (vol 9)

PLATE I

[*Mt. Wilson Observatory*

The spiral nebula (M. 81) in Ursa Major.

A TYPICAL SPIRAL NEBULA.

[*front*

EOS

OR

THE WIDER ASPECTS OF COSMOGONY

By

J. H. JEANS, D.Sc., LL.D., F.R.S.

Secretary of the Royal Society, and Research Associate of Mount Wilson Observatory.

LONDON
KEGAN PAUL, TRENCH, TRUBNER & Co., LTD.
NEW YORK: E. P. DUTTON & Co.
1928

First Published - - *September*, 1928
Second Impression - *November*, 1928

LIST OF ILLUSTRATIONS

LIST OF ILLUSTRATIONS

FOREWORD

This little book contains the Trueman Wood Lecture delivered before the Royal Society of Arts on 7th March, 1928, expanded to more than double its original length by the addition of what I would have liked to say, but had not time to, at the lecture, and by the inclusion of a small part of an earlier lecture " Recent Developments of Cosmical Physics " delivered before the University of London on 9th November, 1926. My thanks are offered to the Royal Society of Arts, and also to the Editor of " Nature " in whose columns the reports of these lectures first appeared, for permission to use them in preparing the present book.

While the " To-day and To-morrow " series has been the medium of publication of many brilliant flights of imagination, the present book is limited to scientific facts and such inferences as I think may properly be drawn from them. Happily astronomy is a science in which exact truth is ever stranger than fiction, in which the imagination ever labours panting and breathless behind the reality, and about which one could hardly be prosaic if one tried.

J. H. JEANS.

DORKING,
12th May, 1928.

EOS

Interest in scientific cosmogony is a recent, and still a very tender growth. Anthropologists and geologists tell us that man has existed on earth for something like 300,000 years; we must go this far back to meet our ape-like ancestry. Between them and us some 10,000 generations of men have walked the earth, most of whom have probably given some thought, in varying degrees, to the significance of their existence and the plan of the universe.

Of these 10,000 generations of men, the first 9990 unhesitatingly regarded the earth as the centre, and terrestrial life as the central fact, of the universe. As was suited to its majesty and dignity as the abode of man, the earth stood still while the celestial sphere spun round it, covering in the earth much as a telescope-dome covers in the telescope; and this dome was spangled with stars, which had been thoughtfully added so as not to leave the central earth unillumined at night. Ten generations at most have been able to view the problem of their existence in anything like its proper astronomical perspective.

The Position of Man in the Universe

The total age of the earth far exceeds the 300,000 years or so of man's existence. The evidence of geology, and of radio-activity in rocks in particular, shows that it must be something like 2000 million years, which is several thousand times the age of the human race. Old Mother Earth must regard man as a very recent apparition indeed; he has just appeared to burrow into her, burn her forests, put her waterfalls into pipes, and generally mar the beauty of her features. If he has done so much in the first few moments of his existence, she may well wonder what is in store for her in the long future ages in which he is destined to labour on her surface. For in all probability the life in front of the human race must enormously exceed the short life behind it. A million million years hence, so far as we can foresee, the sun will probably still be much as now, and the earth will be revolving round it much as now. The year will be a little longer, and the climate quite a lot colder, while the rich accumulated stores of coal, oil, and forest will have long been burnt up; but there is no reason why our descendants should not still people the earth. Perhaps it may be unable to support so large a population as now, and perhaps fewer will desire to live on it. On the other hand, mankind, being three million times as old as now, may—if the conjecture does not distress our pessimists too much—be three million times as wise.

Looked at on the astronomical time-scale, humanity is at the very beginning of its existence—a new-born babe, with all the unexplored potentialities of babyhood; and until the last few moments its interest has been centred, absolutely and exclusively, on its cradle and feeding-bottle. It has just become conscious of the vast world existing outside itself and its cradle; it is learning to focus its eyes on distant objects, and its awakening brain is beginning to wonder, in a vague, dreamy way, what they are and what purpose they serve. Its interest in this external world is not much developed yet, so that the main part of its faculties is still engrossed with the cradle and feeding-bottle, but a little corner of its brain is beginning to wonder.

Taking a very gloomy view of the future of the human race, let us suppose that it can only expect to survive for two thousand million years longer, a period about equal to the past age of the earth. Then, regarded as a being destined to live for three-score years and ten, humanity, although it has been born in a house seventy years old, is itself only three days old. But only in the last few minutes has it become conscious that the whole world does not centre round its cradle and its trappings, and only in the last few ticks of the clock has any adequate conception of the size of the external world dawned upon it. For our clock does not tick seconds, but years; its minutes are the lives of men. A minute and a half ago the distance of a star was first measured and provided a measuring-rod for the universe. Less than a quarter of a minute has elapsed since Professor Hertzsprung of Leiden and Dr. Shapley, now Director of Harvard Observatory, showed how the peculiar stars known as Cepheid variables provide a longer measuring-rod, and taught us to think in distances so great that light takes hundreds of thousands of years to traverse them. With the very last tick of the clock, Dr. Hubble, of Mount Wilson Observatory, using the same measuring-rod, has found that the most remote objects visible in the biggest telescope on earth are so distant that light, travelling 186,000 miles a second, takes about 140 million years to come from them to us.

Not only is our vision of the universe continually expanding, but also it is expanding at an ever-increasing rate. Is this expansion destined to go on for ever? So far as we can at present see, no; for a general guiding principle, that of generalised relativity, fixes a limit, which we are fast approaching. According to this theory, space cannot extend for ever; it has no limit, but is nevertheless finite like the surface of the earth. Without exploring and surveying the whole of the earth's surface, we can make a fair estimate of its total area by measuring its radius, which we can do by measuring its curvature at any one point. In the same way the total volume of space is fixed by a quantity, the curvature of space, which can be determined by measuring the density of distribution of matter in space. Space which contained no matter would go on for ever, but the parts of space we can survey with our telescopes contain enough matter to show that we already see an appreciable fraction of the whole of space. It is as though our baby, watching ships coming from over the horizon, concluded that the earth's surface was curved, and formed a general rough conception of its size by imagining the observed curvature continuing until the earth's surface rounded back on itself.

Exact figures are impossible, but Dr. Hubble has calculated that space is not likely to extend to more than about a thousand times as far as the farthest nebula visible in the biggest telescope. Nothing prevents our going on and on in space beyond this distance, but, if we do, we merely come back to ourselves. The possessor of a sufficiently sensitive wireless apparatus may emit signals and pick them up a seventh of a second later after they have travelled round the world. In the same way a not inconceivable increase in the size of our telescopes would take us round the whole of space, and we should see the stars surrounding our sun by light which had travelled round the universe, not of course as they now are, but as they were 100,000 million years ago.

Such considerations make it improbable that the expansion of the universe can continue at its present rate for much longer. Having grasped that the world is round, the infant speedily forms a fair idea of its size. Our particular infant, mankind, has made the great discovery of the existence of the outer world, has formed some conception of its size, and adjusted his ideas, not by a process of slow revelation, but by a brain-flash of the last few seconds. In his mature years and his staid old age he is no doubt destined to make many sensational discoveries, but he can never again live through the immortal moment at which he first grasped the immensity of the outer world. We only live through a few ticks of his clock, and fate might have ordained that they should be anywhere in the three days that the child has already lived, or in the seventy long, and possibly tedious, years yet to come. The wonderful thing is that she has selected for us what is, perhaps, in some ways the most sensational moment of all in the life of our race.

The child sets its newly awakened mind to work to adjust and co-ordinate a new array of facts. If the world was not made to surround its cradle, what purpose can it serve? If the lights of the great ships in the harbour were not designed to light its nursery at night, what can they possibly be for? And, most interesting problem of all, if the world is such a big affair, can there be other cradles and other babies?

These remarks will have served their purpose if they suggest that the cosmogony of 1928 should not be judged as a finished science or the solution of a problem; it is rather the first confused gropings of the infant mind trying to understand the world outside its cradle. And if the impression produced by its first inexperienced glance at the outer world had to be described in a single word, it would probably select the word "immensity".

THE IMMENSITY OF SPACE

The immensity of space is measured by the figures already mentioned. Light and wireless signals travel at the same rate because, of course, they are essentially the same thing; and this thing takes a seventh of a second to travel round the world, and probably something like 100,000 million years to travel round the universe. The ratio of these times (2×10^{19}) [1] measures the dimensions of

[1] Here, as elsewhere, 2×10^{19} is an abbreviation for 2 followed by 19 zeros, or 20,000,000,000,000,000,000. A million is 10^6, a million million is 10^{12}, and so on. The largest number which occurs in the present book is $10^{420,000,000,000}$. Failing our convenient mathematical shorthand, this number would be expressed by a 1 followed by six million volumes similar to the present, all full of 0's.

PLATE II

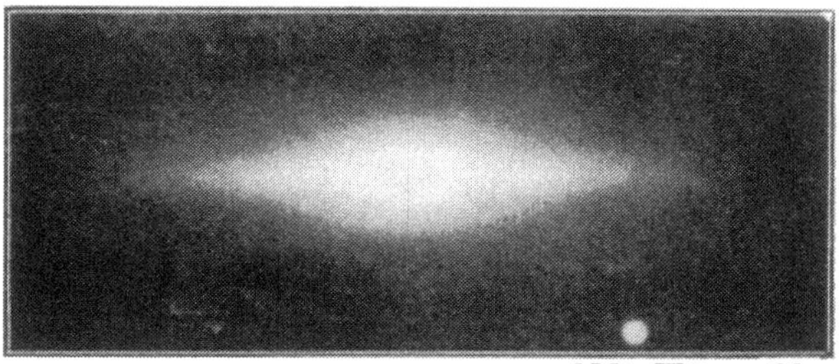

[*Mt. Wilson Observatory*

Fig. 1. Regular shaped nebula (N.G.C.3115).

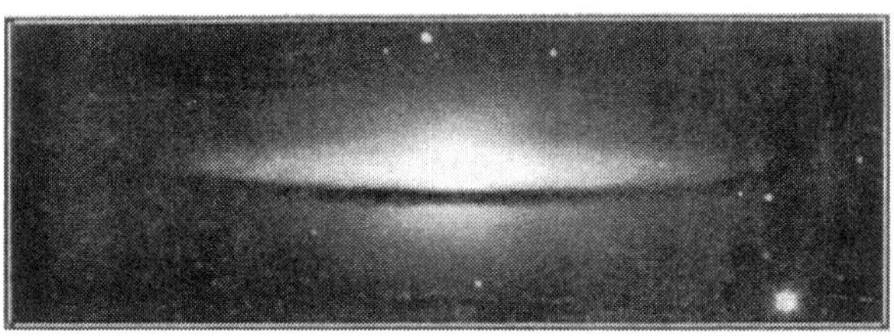

[*Mt. Wilson Observatory*

Fig. 2. Regular shaped nebula (N.G.C.4594) with ring of dark matter surrounding equator.

TWO EARLY STAGES OF NEBULAR DEVELOPMENT.

[*face p. 18*

the universe in terms of the familiar dimensions of the world; incidentally, it also measures the expansion of our spatial ideas since Copernicus. The disparity of size is too great to be easily visualised. Suppose the size of our earth represented by a single atom, whose diameter is about a hundred-millionth part of an inch. Then the range of vision of the biggest telescope is about represented by the whole earth, and the size of the whole universe, according to the theory of relativity, is represented by a stack of a thousand million earths.

Scarcely less bewildering than the immense extent of space is the immense amount and variety of matter it contains. The sun, which is a million times as big as the earth and 300,000 times as massive, proves to be something less than a grain of sand on the seashore. It forms one of a family whose number must certainly be counted in thousands of millions; Dr. Seares has estimated it at thirty thousand millions. This is not the only family of stars in space. Each of the great spiral and other extragalactic nebulæ, such as are shown in the Frontispiece and in Plates II and III is either a family of stars, or consists of stars in the making, or of matter which is in all probability destined ultimately to form stars.

We can estimate the masses of these great nebulæ by gravitational means, and each is found to contain enough matter to make a thousand million suns. This of itself will give some conception of the vast size of these nebulæ, but to tell the whole story, it must be added that their colossal masses are so tenuous that each millionth part of an ounce is, on the average, as big as the Matterhorn. Think of a body which is bigger than the Matterhorn by as much as a thousand million suns is heavier than a millionth part of an ounce, and we have the size of any one of these great nebulæ. Any one of the photographs reproduced in this book would have to

PLATE III

[*Mt. Wilson Observatory*

Fig. 3. Spiral nebula (N.G.C.891) seen edge-on.

A LATER STAGE OF NEBULAR DEVELOPMENT.

[*face*

be enlarged so as to cover at least the whole of Asia before a body of the size of the earth became visible in it at all, even under the most powerful of microscopes.

Dr. Hubble estimates that about two million such nebulæ are visible in the great 100-inch telescope at Mount Wilson, and that the whole universe is about a thousand million times as big as the part of space which is visible in this telescope. Let us now multiply 1000 million by 2 million, and the product by 1000 million. The answer (2×10^{24}) gives some indication of the probable number of stars in the universe; the same number of grains of sand spread over England would make a layer hundreds of yards in depth. Let us reflect that our earth is one millionth part of one such grain of sand, and our mundane affairs, our troubles and our achievements, begin to appear in their correct proportion to the universe as a whole.

The Variety of the Stars

While the stars may be compared to grains of sand in number, they differ too much *inter se* for the comparison to be carried further. They used to be regarded as mere points of light which differed from one another in glory but were too remote for any investigation as to their nature and condition to be practicable. Astronomy can now take their temperatures, by methods similar to those used with factory furnaces, can measure their sizes with a specially-designed stellar interferometer, and can calculate their weights from the gravitational pull they exert on companion stars, just as we calculate the weight of the earth from the pull it exerts on the moon to keep it in its orbit. And the measurements disclose an immense variety of big and little stars, of bright and faint stars, and of hot, hotter and still hotter stars.

The faintest star known, Wolf 359,

emits only a fifty-thousandth part of the light of the sun. If this star were suddenly substituted for the sun, the seas would immediately freeze into solid ice, and the atmosphere condense into liquid air ; people at the Equator would receive rather less light and heat from their new sun than one receives from a coal fire a mile away, and we should meet an icy death. At the other extreme, the most luminous of known stars, S Doradus, emits 300,000 times as much light and heat as the sun. If this star were to replace our sun, our temperatures would run up to about 7,000 degrees Centigrade, which is about twice the temperature of the hottest part of the electric arc, and the solid earth, with its cloud capped towers and gorgeous palaces and ourselves, would immediately dissolve into vapour.

There is almost as much variety in the sizes of the stars. The smallest known star, Van Maanen's star, is about the size of the earth ; a million such stars could be packed inside the sun and leave room to spare. The largest known star, Betelgeux, is so large that 25 million suns could be packed inside it. If we represent the sun by a golf-ball, Van Maanen's star becomes something smaller than the dot of an 'i' on this page, while Betelgeux must be represented by a good-sized house. If Van Maanen's star were to replace our sun, it would appear smaller than Jupiter or Saturn; if Betelgeux were to replace our sun, we should find ourselves inside it, its radius being greater than that of the earth's orbit.

As compared with their enormous ranges of brightness and size, ranges greater than those between a searchlight and a glow-worm, or between balloons and bird-shot, the stars shew a restful uniformity in their weights. No star exists whose weight is known or suspected to be less than about a tenth of that of the sun, while few exist which have as much as ten times the sun's weight—probably about one in a hundred thousand, the other 99,999 all having weights between a tenth and ten times the weight of the sun.

It follows that the different sizes of the stars do not result to any great extent from the different amounts of matter the stars contain; they result rather from the varying closeness or looseness with which the matter is packed. An average ton of matter in the sun occupies about as much space as an ordinary ton of coal in a coal-cellar, but the same amount of matter in Betelgeux takes up as much room as the Albert Hall, while the average ton of matter in Van Maanen's star takes up only as much room as a pea in a pea-pod; a hundred tons of it could be packed into a pocket book with ease. Judged by the standards of solidity which prevail on Van Maanen's star, everything on earth is of the very filmiest of gossamer.

The Structure of the Stars

Their enormous differences of brightness, size and density notwithstanding, the stars are believed to be essentially similar structures.

A normal atom consists of a central nucleus around which a number of electrons revolve in orbital motion like planets round the sun—a miniature solar system, in fact, in which the vacant space far exceeds that occupied by matter. The nucleus, although excessively minute, its diameter being only a small fraction of a millionth of the millionth of an inch, is by no means a simple structure. It consists of a number of particles charged with positive electricity, called protons, and a number of other particles charged with negative electricity, which carry the same charge as, and may be identical with, the electrons which revolve outside the nucleus.

Under the action of great heat the outermost of the atomic electrons begin to break loose from the atom and fly off at a tangent, just as, when water is heated up, the outer molecules break loose and set off on independent journeys of their own. Finally the water is wholly evaporated; the heat has transformed it into a mass of gas (steam) in which each separate molecule flies along its own individual path, like the bullets on a battlefield. In precisely the same way, the application of heat to the atoms causes successive layers of electrons to break loose from their moorings, and the atoms become smaller and smaller until finally no coherent structure remains but merely a powdered *débris* of atomic constituents, each nucleus and electron going its own way regardless of the rest.

We cannot study the process to any extent in the laboratory since we cannot command high enough temperatures. It is just beginning in hot flames and in the electric arc. The atmospheres of stars are at higher temperatures than any obtainable on earth, and here the process is further advanced; the spectroscope shews that the atoms may be broken into two, three or even four pieces. The temperatures in the interiors of the stars are higher still, and although we cannot measure them directly we can calculate them with very fair precision. The temperature at the centre of the sun is found to be in the neighbourhood of 50,000,000 degrees, and this is a fairly average temperature for the stars in general. It is difficult to form any conception of the physical meaning of a temperature which is so far removed from anything in our experience. If a piece of matter the size of a pea could be maintained at this temperature it would scorch and shrivel up anyone who ventured within a thousand miles of it; its directed beam would destroy an army in an instant. Nevertheless, the mathematician need not hesitate to thrust his calculations right into the very hearts of the stars, and he can shew, with something approaching very near to certainty, that at the centres of most of the stars nearly all, or perhaps quite all, of the electrons must have broken loose from their parent atoms, leaving the stellar matter almost or quite pulverized into its constituent nuclei and electrons.

When I first put forward this view in 1917, I thought it was entirely novel, but I have since found that in 1644 Descartes had conjectured that the sun and fixed stars were made of matter "which possesses such violence of agitation that, impinging upon other bodies, it gets divided into indefinitely minute particles". My own suggestion was not conjecture, but was an inevitable deduction from modern knowledge of atomic physics. Since it was put forward, a great deal of labour has been devoted to testing the hypothesis that practically all the electrons have broken loose from their atoms, the stripped atoms and electrons flying about in a general hurly-burly like the molecules of a gas. But the hypothesis has proved disappointing, and a much more probable hypothesis is, I think, that the atoms are not stripped quite bare, but that in most stars they retain a few rings of electrons which give the atoms so much size that they jostle one another about like the molecules of a liquid. This hypothesis explains beautifully the otherwise puzzling fact that the stars tend to fall into distinct groups, of what may almost be described as 'standardised' sizes. On the 'liquid star' hypothesis, these different sizes correspond to the different sizes possible for the stellar atons, which may have 0, 1, 2, or 3 rings of electrons left, but cannot have fractional numbers. The largest stars of all, such as Betelgeux, have three rings left, while minute stars, such as Van Maanen's star, consist of atoms most of which are stripped quite bare, so that there is almost no limit to the closeness with which they can be packed together. These stars of course represent extremes of size; more normal stars such as Sirius, Procyon and the Sun consist mainly of atoms in which one ring of electrons is left revolving about the nucleus.

Thus the observed sizes of the stars proclaim the secret of the structure of the atom. The sizes of the stars are discontinuous because the sizes of atoms broken down to different stages are discontinuous. These discontinuities can be traced in turn to the discontinuities which form the central feature of the new system of quantum dynamics, which is now known to dominate the whole of atomic physics. Thus the distinguishing characteristic of the laws which govern the most minute processes in Nature is transmitted directly into the large scale phenomena of astronomy and governs the distribution of the huge masses of the stars. The infinitely great is never very far from the infinitely small in science, but it would be hard to find a more sensational illustration of the unity of science than that just given.

On this hypothesis, not only do the observed sizes of the stars disclose the general structure of the atom, which is old knowedge, but they also reveal the detailed structure of the particular atoms of which the stars are composed, and this is new knowledge. The more complex the structure of an atom is, the higher the temperature needed to break it up. We find that simple atoms would be completely broken up in the interiors of the stars; no coherent atomic structure would remain. To retain the various sizes demanded by the theory of liquid stars, the stellar atoms must be heavier and more complex than any atoms known on earth. The atoms which reveal their presence in the spectra of the stars are, of course, atoms of the ordinary terrestrial elements—hydrogen, iron, calcium, and the like. These, being the lightest atoms in the star, must naturally float up to its surface and so determine its spectrum. Moreover, as the earth was originally formed out of the surface of the sun, the earth is necessarily composed of the same light elements. But it now appears likely that down in the depths of the stars are other unknown and heavier atoms. Later on, we shall find another reason why this must be so, namely that no terrestrial atoms, not even radium or uranium, can produce anything like the amount of energy which these stellar atoms are observed to produce.

The Immensity of Time

The immensity of space is paralleled by that of time. We can estimate the ages of stars from the impression that time has made upon them, just as we estimate the age of a tree from the number of subdivisions of its stem, or of rings in its cross section. There are three principal methods of doing this. The orbits of binary stars, which are circular at birth, are gradually knocked out of shape by the forces from passing stars. As we can calculate the rate at which this process occurs, the shape of stars' orbits can be made to reveal their ages. The moving clusters provide a second method. Groups of bright stars such as the Great Bear, the Pleiades, Orion's Belt, are often found to consist of exceptionally massive stars which move in regular orderly formation through a jumble of slighter stars, like a flight of swans through a confused crowd of rooks and starlings. Swans, however, are conscious beings, and continually adjust their flight so as to preserve their formation. The swan-like stars cannot do this, so that their orderly formation must in time be broken by the gravitational pull of other stars. When this happens, the lighter stars are naturally knocked out of formation first, while the most massive stars retain their formation longest. This agrees with what is observed, and as we can calculate the time necessary to knock out the lighter stars, we can at once deduce the ages of those which are left in. A third method of investigation rests upon a rather abstruse dynamical theorem, which shows that after a sufficient time the energies of motion of the different types of stars must tend to equality, the little stars making up for the smallness of their mass by the rapidity of their motion. Dr. Seares, of Mount Wilson Observatory, has shown that the stars near the sun have nearly attained to this ideal state, and as we can calculate the time needed to establish it, we can again deduce the ages of the stars.

It is gratifying and significant that all three lines of investigation lead to the same result: the stars are found to be some millions of millions of years old, perhaps from five to ten millions of millions. We cannot state their ages with much precision, but it is the general order of magnitude, not the exact figure, that is important.

The Source of Stellar Radiation

Year after year, century after century, for millions of millions of years, the sun radiates enough energy from each square inch of its surface to keep a fifty horse-power engine continually in action; still hotter stars may radiate as much as 30,000 horse-power per square inch. If this energy were produced by the combustion of coal, the stars would all be completely burnt out in a few hundreds or thousands of years. Where, then, shall we find a source of energy to last millions of millions of years?

More than twenty years ago I directed attention to the enormous store of energy made available by the annihilation of matter, by positively and negatively charged protons and electrons falling into and annihilating one another, thus setting free the whole of their intrinsic energy as radiation. On this scheme neither energy nor matter had a permanent existence, but only a sort of sum of the two; each was, theoretically at least, convertible into the other.

When I put forward this hypothesis, I thought I was advocating something entirely revolutionary and unheard-of, but I have since found that Newton had anticipated something very similar exactly two centuries earlier. In his *Opticks* (1704) we find:

"*Query* 30. Are not gross bodies and "light convertible into one another; and "may not bodies receive much of their "activity from the particles of light which "enter into their composition?

"The changing of bodies into light, and "light into bodies, is very conformable "to the course of Nature, which seems "delighted with transmutations. Water, "which is a very fluid, tasteless salt, she "changes by heat into vapour, which is "a sort of air; and by cold into ice, which "is hard, pellucid, brittle, fusible stone; "and this stone returns into water by heat, "and vapour returns into water by cold. . . "Eggs grow from insensible magnitudes, "and change into animals; tadpoles, into "frogs; and worms, into flies. All birds, "beasts and fishes, insects, trees, and "other vegetables, with their several parts, "grow out of water and watery tinctures "and salts; and by putrefaction, return "again into watery substances. And water, "standing a few days in the open air, "yields a tincture, which (like that of "malt) by standing longer yields a sediment "and a spirit; but before putrefaction is fit "nourishment for animals and vegetables. "And among such various and strange "transmutations, why may not Nature "change bodies into light, and light "into bodies?"

For reasons which will appear later, it is highly improbable that light is ever transformed into matter, but the falling together of positively charged atomic nuclei and negatively charged electrons provides an obvious mechanism for the transformation of matter into energy, and, as we shall now see, it is practically certain that this is the actual source of the radiation of the stars.

A beam of radiation exerts pressure on any surface it falls upon, just as a jet of water does or a blast of air. The reason is that radiation carries mass about with it, and electromagnetic theory tells us the amount of this mass. For example, we can calculate that a searchlight which is radiating 50 horse-power of energy is discharging mass into space with the radiation at the rate of a gramme and a quarter a century; with sufficiently delicate adjustments it might even be possible to observe the recoil of the searchlight. Indeed, the pressure of radiation has actually been measured, although not in this particular way. New mass is of course being continually fed into the searchlight by the electric current; if it were not so, the searchlight would weigh a gramme and a quarter less after functioning for a century, this being the weight of the light it had emitted in this period of time.

Each square inch of the sun's surface is in effect a searchlight discharging radiation into space at the rate of 50 horse-power, and so is discharging mass at the rate of a gramme and a quarter a century, and the sun's surface is so large that the sun as a whole is discharging mass into space at the rate of 250 million tons a minute. Now the sun has no source of replenishment. It must have weighed 360,000 million tons more yesterday than to-day, and by to-morrow will weigh 360,000 million tons less. These are not mere speculative statements; they rest on observation, and on generally accepted principles which are directly confirmed by observation.

Allowing for the fact that a more massive star emits more radiation than a less massive one, we can calculate that five or ten million million years ago the sun must have been several times as massive as it is to-day, so that it has already lost most of the mass it had at birth. Of each ton it had at birth only a few hundredweights at most remain to-day. The loss of mass which accompanies radiation is, then, no mere academic hair-splitting. It is a real astronomical phenomenon, and young stars must be many times as massive as old stars.

There is a certain amount of direct evidence of this change of mass. The radiation of the stars imposes an endlessly recurring capital levy upon their masses, which, as observation shows, is graduated and increases very steeply indeed for the richest stars. The levy makes all the stars poorer, but it also tends to equalise what wealth remains; the older the stars get, the more nearly equal their impoverished masses become. This is a large part of the reason why the stars are nearly equal in mass. The process is most clearly marked in the binary systems, which have been formed by a single star breaking into two. The two component stars of such a system are necessarily of the same age, and it is a matter of observation that the small stars of old systems are nearer to equality of mass than the massive stars of young systems.

Thus observation and theory agree in indicating that the universe is melting away into radiation. Our position is that of polar bears on an iceberg that has broken loose from the icepack surrounding the pole, and is inexorably melting away as the iceberg drifts to warmer latitudes and ultimate extinction.

The Annihilation of Matter

Five million million years ago the sun had stored up within itself the energy which was destined to provide its light and heat until to-day, and the mass of this energy was many times the present mass of the sun. No means is known by which so much mass could be stored except in the form of electrons and protons. Thus we must suppose that the radiation of the sun through these millions of millions of years has been produced by the annihilation of electrons and protons which existed in it originally, but no longer exist now. These electrons and protons are pure bottled energy; the continuous breakage of these bottles in the sun sets free the radiation which warms and lights our earth, and enough unbroken bottles remain to provide light and heat for millions of millions of years to come.

The amount of energy made available in this way is amazing. The annihilation of a pound of coal a week would produce as much energy as the combustion of the five million tons a week which are mined in the British Isles; an ounce of coal a month would provide locomotive power for all the British railways, while a single drop of oil would take the *Mauretania* across the Atlantic. When we speak of the efficiency of a steam engine as 5 per cent. or so, we regard complete use of the thermal energy of combustion as 100 per cent. efficiency. If we measure the work done against the total intrinsic energy of the fuel, as made available by its complete annihilation, the efficiency is more like 0·00000001 per cent. On this scale the efficiency of the sun and stars is exactly 100·00 per cent.

Modern physical theory shows that the annihilation of an electron must produce a single flash of radiation of wave-length far shorter than any we can produce on earth. As this radiation threads its way through a star, its wave-length is continually increased, or, to use the technical term, the radiation is continually softened. In time it becomes γ-radiation, then hard X-radiation, then soft X-radiation, and finally it emerges from the surface of the star as ordinary light and heat. Consider, however, an electron which is annihilated not inside a star but outside in free space, or in one of the almost transparent nebulæ. The short wave-length radiation now undergoes no softening, but travels on until it meets something capable of checking it Thus all astronomical bodies, including the surface of the earth, ought to be under continual bombardment by radiation of shorter wave-length, and consequently of greater penetrating powers, than any we can produce on earth.

Highly Penetrating Radiation

Many years ago such radiation was detected in the earth's atmosphere by Prof. McLennan, now of Toronto, by Sir Ernest Rutherford, and other observers; it has recently been studied in detail by Prof. Millikan of Pasadena and others. There is no reason to doubt that it originates just where it ought to, namely in the great nebulæ, and its amount is about what it ought to be, if it is evidence of the whole universe melting away into radiation. The wave-length of the radiation might be expected to reveal the physical process by which it is generated, but the evidence is a bit puzzling. The hardest terrestrial radiation penetrates inches of lead and corresponds to a voltage of hundreds of thousands of volts. The cosmic radiation penetrates about five yards of lead, and the hardest rays are now found to correspond to about 60 million volts. Prof. Millikan was at one time inclined to attribute the rays to the combination of four atoms of hydrogen to form an atom of helium, but rays so produced would only be of the hardness corresponding to 30 million volts. There are many ways known to physics of softening radiation, but none of hardening it. Thus we must look for some source more energetic than the synthesis of hydrogen into helium, and I can see no possible stopping-place short of the annihilation of matter. Again, we are not dealing with a minute phenomenon of mere academic interest. In a sense this radiation is the most fundamental physical phenomenon of the whole universe, most regions of space containing more of it than of visible light or heat. Our bodies are traversed by it night and day. Short of going down into a mine or in a submarine we cannot escape it, and it is so intense that it breaks up several million atoms in each of our bodies every second. It may be essential to life or it may be killing us.

The Final End of Radiation

There has been much discussion as to the ultimate fate of astronomical radiation. An almost infinitesimal fraction strikes the earth in the various forms of sunlight, starlight and highly-penetrating radiation, and further minute fractions must similarly strike, and be absorbed by, other bodies in space, but the vast majority will merely wander on and on without ever meeting a target of any kind. So long as space was believed to be infinite in extent, radiation could be imagined to travel for ever onwards, but in a finite universe it can only travel round and round like a squirrel in a cage. It has been suggested that the radiation poured out from millions of millions of stars through millions of millions of years may ultimately cause space to become overcrowded with radiation, just as a cage would become overcrowded with squirrels if we kept putting them in and never took any out. Or again, it may be as though we kept putting lump after lump of sugar into the same cup of tea until finally the tea became saturated with sugar and refused to dissolve any more, at any rate unless some of the sugar already in solution crystallised out afresh as new sugar.

Calculation shews that the universe is in no danger of approaching either of these conditions. Let us pass at once to the extreme limit and imagine every atom of the present universe, stars, nebulæ and all else, suddenly transformed into radiation. We can estimate how much matter there is in the average cubic mile of the present universe, and so can calculate how much radiation there would be to the cubic mile if this matter all dissolved into radiation. We find that it would only raise the temperature of space from absolute zero to about eleven degrees above absolute zero, or to —262 degrees Centigrade, a temperature well below that of liquid air. It is known that the temperature of the earth's surface is determined almost entirely by the amounts of heat and light it receives from the sun, the radiation from all other stars, including that wandering round and round space, contributing practically nothing. We can calculate that if the whole universe, apart from the sun, were suddenly to melt into radiation, the temperature of the earth would be raised less than a five-thousandth part of a degree, and, except for the stars being blotted out, we should notice absolutely nothing. The radiation of ten thousand dead universes may even now, for ought we know, be wandering round space; nothing less than the radiation from hundreds of thousands would be susceptible of scientific measurement. Thus we cannot tell how many universes may have perished into radiation and be wandering in this ghostly form around space, but there is certainly ample room for one more.

With an ardour equalled only by that of man's longing for personal immortality, many seem to desire that the universe itself should prove in some way to be immortal. Science sees matter being annihilated here, they argue, and its energy transformed into radiation, but in other parts of the universe this very radiation may be forming new heavens and a new earth; the tide only ebbs on our shores because it is rising on some other distant shore unknown to us.

The considerations we have just mentioned are sufficient to dispose of any such belief, but the matter can be further tested by direct numerical calculation. We have seen that if space starts empty of all radiation the melting away of our present universe will raise its temperature by about 11 degress. And calculation shews that the universe will not become saturated with radiation, in the way in which the tea may become saturated with sugar, until it reaches a temperature of 7,500,000,000,000 degrees. At this temperature, and not before, space refuses to absorb any more radiation, or if it does, it rids itself of an equivalent amount by forming new matter out of it. And to raise space to this temperature would require the dissolution of some 3×10^{47} universes similar to our present universe. The number is millions of times greater than the number of grains of sugar that could be melted in the Atlantic before this began to crystallise out into new sugar. Thus the melting away of the whole of the present universe into radiation is something less than the melting of a single grain of sugar in the Atlantic, and the radiation of the stars is as far from creating new worlds as is the sweetness of a single grain of sugar from turning an ocean into candy. Space is so vast by comparison with the amount of matter it contains that our simile of melting lumps of sugar in a tea-cup was wrongly conceived; we ought rather to have discussed the dropping of a single grain of sugar into the Atlantic.

We may sum up by saying that the

capacity of space for radiation is practically infinite when judged by any amount of radiation which can ever be poured into it. It follows that the transformation of matter into radiation is a "one-way", or as it is technically called, an "irreversible" process. Matter can change into radiation, but under present conditions radiation can never change back into matter. Ultimately a time must come when every atom which is capable of dissolving into radiation will have done so. The universe is like a clock which is running down, a clock which, so far as science knows, no one ever winds up, which cannot wind itself up, and so must stop in time. It is at present a partially wound-up clock, which must, at some time in the past have been wound up in some manner unknown to us. By studying the mechanism of the clock, and noting the length of spring which is still coiled up and the length already uncoiled, we can estimate the length of time the clock has still to run, and the length of time since it was wound up, but we can obtain no evidence as to the way in which it was originally wound up and set going.

Our scientific ancestors of half a century ago were wont to regard the universe as a fortuitous concourse of atoms which, created they knew not when or how, had fallen together and chanced to form the earth and the starry heavens. The wider knowledge of to-day shews that the main mass and the main energy of the universe do not exist in the form of atoms but of intangible radiation. We may say that the universe is mainly a universe of radiation, combined, in a far lesser degree, with the atoms out of which radiation is continually being formed. Can we regard this new universe as a fortuitous concourse of atoms and radiation ?

The analogy of the clock at once suggests a negative answer, which exact calculation emphatically confirms. A fortuitous concourse of any ingredients whatever must after sufficient time reach a state which the physicist describes as a " condition of maximum entropy ", but which for the sake of brevity we may call the " final state " ; it is the state of the clock which has run completely down. No matter how the ingredients start, provided they are always the same ingredients, their final state is always the same. If the ingredients are a cup of tea and four lumps of sugar, the final state is a uniform mass of sticky liquid of rather sickly sweetness, and this is always just the same, no matter how or where or in what order I put the four lumps of sugar in my tea.

The final state of the universe admits of very exact calculation. We only need to know the ingredients, which are matter and radiation ; we do not need to know either their present arrangement or distribution, or the mechanism by which the universe works towards its final state. All roads lead to Rome and Rome is very easily located. Calculation shows that in the final state the odds against even a single atom which is capable of dissolving into radiation remaining undissolved are quite unthinkably great, being something like $10^{420,000,000,000}$ to one. Since many such atoms are still undissolved, we may dismiss the possibility of the present universe being in its final state. Now the odds against the present division of the total energy of the universe into atoms and radiation being fortuitous are, as it happens, precisely the same as the odds against the universe having reached its final state; indeed the mathematical specification of a fortuitous state is precisely the same as that of a final state, and this enables us to dismiss the fortuitous conception of the universe as being entirely out of the question. Everything points with overwhelming force to a definite event, or series of events, of creation at some time or times, not infinitely remote. The universe cannot have originated by chance out of its present ingredients, and neither can it have been always the same as now. For in either of these events no atoms would be left save such as are incapable of dissolving into radiation; there would be neither sunlight nor starlight but only a cool glow of radiation uniformly diffused through space. This is, indeed, so far as present-day science can see, the final end towards which all creation moves, and at which it must, at long last, arrive.

> "Then star nor sun shall waken,
> Nor any change of light;
> Nor sound of waters shaken,
> Nor any sound or sight: . . .
> Only the sleep eternal
> In an eternal night."

The Evolution of Stellar Matter

Let us now turn our attention from stellar radiation to the stellar atoms which produce this radiation by self-annihilation. The fundamental process by which radiation is released is the falling into one another of a pair of oppositely charged electric particles. When this happens the particles annihilate one another and disappear. In their place appears a splash of radiant energy which travels through the star until, after innumerable absorptions and re-emissions, it reaches the star's surface and wanders off into space. Each splash is similar to the splashes produced by radio-active material in the spinthariscope, except for being many thousands of times more powerful. The great energy of the splashes is to some extent counterbalanced by their rarity. In the sun, for example, only about one atom in every 10^{17} annihilates itself each hour. A cubic inch of the sun's mass contains, let us say, 10^{23} atoms, and of these about 1,000,000 are annihilated every hour. The energy produced in a cubic inch of the sun's mass is thus not very great, averaging about 150,000 ergs per hour, which is roughly the energy of a hundred-thousandth part of a candle-power; the enormous flow of energy from the sun's surface results from the fact that all the energy produced in a cone 433,000 miles in depth has to stream out through the mouth of this cone.

We naturally inquire what types of atoms can be responsible for the generation of this radiation. The sun's present radiation is such as would be produced by an average generation of two ergs per second for each gramme of its mass, and to the best of our knowledge it has generated and radiated at this, or a greater, rate, for some millions of millions of years. Could the sun have any such radiating capacity if its interior were formed of the common terrestrial elements, hydrogen, oxygen, calcium, iron, silicon, etc.?

One's first impulse is to say, No. Even if the sun were built of pure uranium, its radiating power would be only about one-half of that observed, and would only last for a minute fraction of what is believed to have been the sun's life. A sun of pure radium would radiate more than enough for the moment, but its life would be limited to a few thousand years. No possible combination of terrestrial elements can give the combination of high radiation and of staying power which is observed in the sun and stars.

We must, however, remember that stellar interiors are at pressures and temperatures which are quite unattainable in our laboratories. We are led to wonder whether our terrestrial elements would behave quite differently if they were exposed to stellar conditions. Is it possible, for example, that the sun's interior is formed of ordinary terrestrial elements, which owe their high generation of energy merely to their high temperatures and pressures? Or, to put the question in a more general form, does the generation of energy proceed more merrily, do the electrons and protons annihilate one another more frequently, when the stellar matter is in a state of high temperature and pressure? Are atomic suicides more frequent when the heat and pressure are greater?

General theoretical physics frowns on any such suggestion, while dynamical theory demurs to it on the grounds that it would probably cause the whole universe to turn explosively into radiation. But perhaps the most direct argument against the suggestion is provided by observational astronomy.

The stars which are observed to radiate most energetically are not, broadly speaking, those in which the stellar material is at the highest temperatures and pressures. Indeed such stars are often enough put to shame in the matter of radiation by cool stars in which the pressure is low. The star S. Doradus, which we have mentioned as being the most luminous of known stars, is also one of the coolest and its interior is at an exceptionally low pressure, while Wolf 359, the least luminous of known stars, has stupendous temperatures and pressures in its interior. If we arrange the stars in order of their energy of radiation per unit mass, we shall find we have arranged them neither in order of temperature nor of pressure, but in order of their weights; the most massive stars radiate most energetically. And, as the life history of a star consists primarily of a continuous loss of weight as its substance melts away into radiation, the heaviest stars are also the youngest and the least massive stars are the oldest. We accordingly find that the stars which radiate most energetically in proportion to their masses are the youngest stars, regardless of their interior temperatures and pressures; the older stars appear to be tired out.

Now if the radiation of the stars emanated from ordinary terrestrial matter raised to high temperatures and pressures we should expect to find those stars in which the temperatures and pressures were highest radiating the most energetically. This is emphatically not observed. Instead we find that energetic radiation is an accompaniment of stellar youth. We conclude that radiation proceeds from types of matter which annihilate themselves, and so disappear altogether, as the star ages. We cannot say for certain that our terrestrial atoms contribute nothing at all, but they can at best contribute very little. If the average atom of the earth generated even a ten-thousandth part as much energy as the average atom of the sun, the earth's surface would be too hot for human habitation. Our terrestrial atoms have so little capacity for annihilating themselves and forming radiation that they may properly be described as "permanent". The atoms of radioactive elements of course form an exception; they probably represent the last surviving vestiges of more vigorous primeval matter, thus forming a bridge between the inert permanent atoms and the more short-lived atoms which generate the radiation of the stars.

Lucid Matter

As far back as 1692 Newton, writing to Dr. Bentley, postulated that the stars must be formed of a special "lucid" type of matter, different from ordinary terrestrial or "non-lucid" matter. As we have seen, a study of modern astronomy shews that the distinction was a sound one. It is still essential to an understanding of the main mechanism of the universe. The distinction is not quite so clear-cut as Newton thought, since the radioactive atoms probably provide a sort of bridge or transition between the two. The lucid atoms are probably the "parents" of the radioactive atoms, just as the latter are, at least in part, the "parents" of the permanent atoms. The scheme to which we are led may be indicated diagrammatically as follows :—

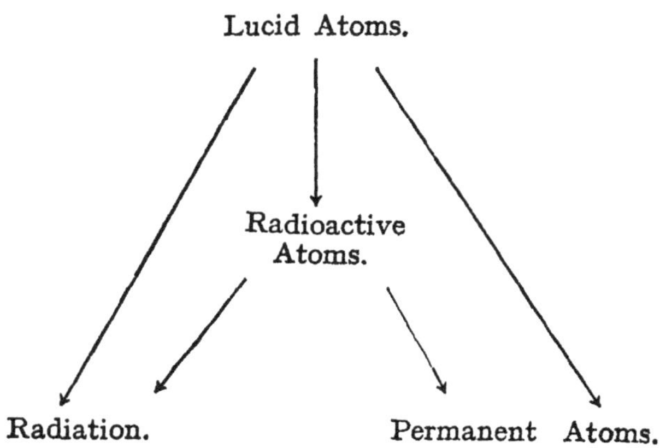

There are probably 92 types of atoms existing on earth, although two of these have not yet been identified or isolated. They are distinguished by the number of electrons which revolve about the central nucleus, this number being commonly called the "atomic number". It ranges from 1 in hydrogen, which is the lightest and simplest atom possible, up to 92 in uranium, the heaviest and most complex of all atoms known on earth. It is significant that all the atoms with atomic numbers below 84 are 'permanent', shewing no tendency to change into simpler elements or radiation, while all with atomic numbers above 83 are 'radioactive', continually changing into simpler atoms and into radiation, and so disappearing from the earth even as we write and read.

As we have been led to look on the radioactive elements as being a half-way house between the 'permanent' atoms of the earth and the 'lucid' atoms of the stars, it is natural to expect these latter to have atomic numbers higher than 92; indeed there is hardly any alternative open to us, since all but two of the first 92 places are already known to be occupied by terrestrial atoms. This fits in exactly with the requirements of the theory of liquid stars, which, as we have already seen, requires stellar atoms to be just a bit heavier and more complex than uranium, so as to resist being completely broken up by the intense heat of stellar interiors.

The lucid atoms being the heaviest atoms in a star, must naturally sink to its centre, while the permanent atoms, being the lightest, float to the surface. Thus the atoms which reveal their presence in the spectre of the sun and stars, in the light emitted by their outermost surfaces, can be only the lightest of the atoms existing in the star. The earth, formed originally out of the sun's outer layers, can only contain the lightest of the atoms existing in the sun. This is why the earth contains only permanent atoms which do not dissolve into radiation. This circumstance alone makes life possible on earth, for if the earth contained any appreciable proportion of lucid atoms it would be too hot for us to live on. The same circumstance endows the earth with a kind of melancholy immortality; it is exempt from the general destiny of matter to turn into radiation, and will continue to exist long after the stars have turned into darkness and all life and light have vanished from its surface.

We have to suppose that primeval matter consists of a mixture of atoms of varying atomic weights and numbers, and so of varying degrees of complexity. The lucid elements, which are the most complex, are the first to disappear, their disappearance reducing both the star's capacity for radiation and also the average complexity of its atoms. Just as, on the coast, the hardest rocks survive for longest the disintegrating action of the sea, so, in a star, the lightest element survive for longest the disintegrating action of time. Ultimately the star loses all radiating power and consists only of the lightest atoms of all, the permanent atoms.

Thus the general trend of chemical evolution in the universe as a whole appears to be from complex to simple, radiation being regarded as the simplest entity of all. This is contrary to the views of the early spectroscopists, who believed they had found evidence of a progressive change from simple to complex. They found, for example, that the spectrum of Sirius exhibited hydrogen lines very strongly and calcium lines very weakly ; in the solar spectrum the relative strength of these two sets of lines was reversed, calcium being strong and hydrogen weak. They concluded that hydrogen was specially prominent in the constitution of Sirius and calcium in that of the sun. Believing that Sirius must one day develop into a star similar to our sun, they conjectured that its substance must gradually change from hydrogen into calcium and other more complex elements, thus finding support for the long-established hypothesis that the more complex elements were formed by gradual evolution out of the simplest.

The true interpretation of these early observations, as the recent investigations of Saha, R. H. Fowler and Milne have abundantly proved, is merely that the surface of Sirius is at a temperature at which hydrogen is specially active in emitting and absorbing radiation, while the sun's surface is at a lower temperature at which hydrogen is comparatively inert, calcium, iron, etc., having become active in its place. Just as the laboratory physicist can produce different spectra from the same vacuum-tube by varying the mode and conditions of excitation, so Nature produces different spectra from the same stellar material by varying its temperature.

Clearly this circumstance robs stellar spectra of all direct evolutionary significance. The spectra of the stars merely tell us their present surface temperatures, so that, even if we could arrange the stars in order of age, a comparison of their spectra would only show whether their surfaces were becoming hotter or cooler; it would give no information as to evolutionary chemical changes occurring in their substance.

To investigate such changes we must probe down below the surface layers from which the star's spectrum emanates, and a study of the fundamental changes in the main body of the star suggests, as we have seen, an evolution from complex to simple—from the complex lucid atoms to simpler atoms and radiation. Such chemical evolution as we are acquainted with on earth, that of the radioactive elements, is in the same direction. This is the opposite direction to biological evolution, which proceeds from simple to complex. If the inanimate universe moves in the direction we suppose, biological evolution moves like a sailor who runs up the rigging in a sinking ship.

Although the range in the masses of the stars is not very great, it is far greater than can be attributed to variation in the atomic weights of the atoms of which they are composed. A very massive star such as Betelgeux, which has perhaps fifty times the weight of the sun, must contain more atoms than the sun. And, as we suppose that Betelgeux will in time shrink into a star similar to the sun, it follows that this shrinkage must be accompanied by an actual annihilation of atoms. There will be fewer atoms in Betelgeux in the future than now. It is not enough to think of the complex atoms breaking down into simpler atoms as the radioactive atoms do; they are destined to disappear completely.

Yet any 'permanent' atoms there may now be in Betelgeux must necessarily survive in the final shrunken star of perhaps only one per cent. of the mass of the present star, so that ninety-nine per cent. of the present mass must consist of non-permanent atoms. The primeval matter of the universe must then consist mainly of non-permanent atoms; atoms of our terrestrial type are the exception rather than the rule. They are a mere by-product of the main processes of the universe, a residue of ash which persists merely because it is incapable of combustion. We begin to realise that our terrestrial physics and chemistry are only the outermost fringes of far-reaching sciences. Beyond the seashore we have explored in our laboratories lies the ocean whose existence we are just beginning to suspect.

The need for caution in our interpretation of the universe is borne in upon us with overwhelming force when we reflect that the new world in which astronomy moves to-day is all a discovery of the present century. It is not merely that our present concepts of the extent of the universe in space and of its duration in time are new revelations to us; our understanding of its fundamental mechanism is equally new. The conversion of matter into radiation, which appears to be the primary physical process of the universe, did not come within our terrestrial purview at all until 1904. The primary matter of the universe appears to consist of elements whose existence we are only just beginning to suspect, and to exist in the state of almost completely broken-up atoms, a state of matter which, again, was not contemplated before 1917. The primary radiation of the universe is not visible light, but short-wave radiation of a hardness which would have seemed incredible at the beginning of the present century. Indeed, our whole knowledge of the really fundamental physical conditions of the universe in which we live is a growth of the last quarter of a century.

The simple explanation of this situation is to be found in the fact that life, naturally enough, begins its exploration of Nature by studying the conditions which immediately surround it; the study of the general conditions of the universe as a whole is a far more difficult task which life on this planet is only now approaching. Now the physical conditions under which life is possible form only a tiny fraction of the range of physical conditions which prevail in the universe as a whole. The very concept of life implies duration in time; there can be no life where the atoms change their make-up millions of times a second and no pair of atoms can ever become joined together. It also implies a certain mobility in space, and these two implications restrict life to the small range of physical conditions in which the liquid state is possible. Our survey of the universe has shown how small this range is in comparison with the range of the whole universe. Primeval matter must go on transforming itself into radiation for millions of millions of years to produce an infinitesimal amount of the inert ash on which life can exist. Even then, this residue of ash must not be too hot or too cold, or life will be impossible, and endless other conditions, of which we know nothing, must in all probability be satisfied before life comes into being. We cannot be too suspicious of the interpretation which our minds, trammelled with long brooding over, and experience limited to, one tiny corner, puts on the greater universe we are just beginning to discover.

The Lives of the Stars

The stars are almost certainly born in nebulæ of the type of the great extra-galactic nebulæ, such as are shown in Plates I, II, III, and IV. These nebulæ show a great variety of shapes, but a single thread connects them all; they are the shapes of huge masses of gas endowed with different amounts of rotation. So definitely is this the case that when Dr. Hubble recently tried to classify the shapes of these nebulæ, deliberately and avowedly shutting his eyes to all theoretical considerations, he found that purely observational considerations compelled him to classify them in precisely

PLATE IV

[Mt. Wilson Observatory

Fig. 4. The spiral nebula (N.G.C.7217).

A SPIRAL NEBULA VIEWED FULL-ON.

[face

the sequence I had predicted on theoretical grounds some ten years earlier.

A huge mass of gas which was entirely devoid of rotation would of course assume a strictly spherical shape ; rotation would flatten this shape out, just as the earth is flattened by its rotation, until ultimately most of the matter was spread out in a thin disc. We see the process beginning in Fig. 1 (p.18), it is well advanced in Fig. 2, and has assumed a rather extreme form in Fig. 3 (p. 20). Fig. 4 shows a nebula which is probably physically similar to that shown in Fig. 3, but viewed from another angle. The nebula shewn in the frontispiece is again probably similar but even more advanced. Now mathematical theory shows that the thin disc-like structure could not remain a mere featureless mass of gas. Just as the cooling of a cloud of steam causes it to condense into drops of water, so the cooling of a cloud of gas causes it to condense into detached masses. We see the phenomenon in progress in any collection of nebular photographs, such as those shewn in the present book; it is a necessary theoretical consequence of the laws of gases and the law of gravitation.

Now the same theory which predicts that the phenomenon must happen, predicts the scale on which it will happen. We can calculate how much matter will go to the formation of each 'drop', and the calculated masses of the drops come out to be just about the same as the masses of the stars. Indeed these drops are stars, and the process just described is that of the birth of stars. Unmistakable stars have been observed in the outer regions of many of the spiral nebulæ, including the nebula M 81 which figures as the frontispiece. It is naturally not possible to identify every observed spot of light with a star, but some of them show precisely the same peculiar fluctuations of light as characterise a certain class of variable star, the Cepheid variables already mentioned, and these put the

PLATE V

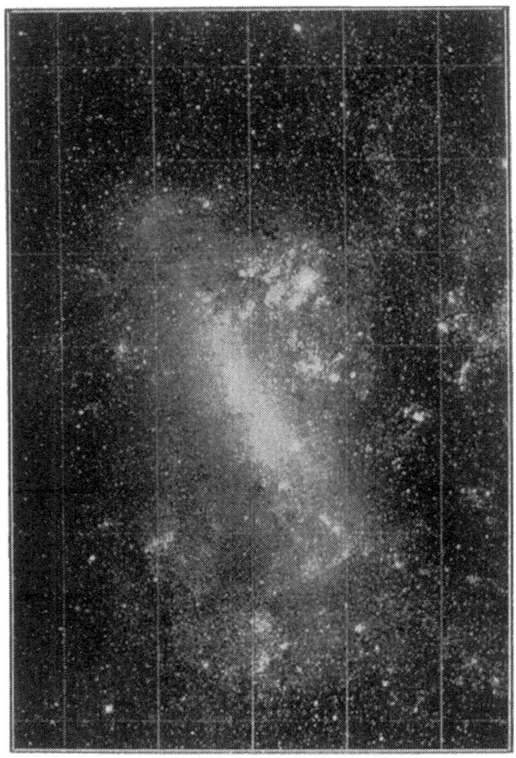

[*Franklin-Adams Chart*

Fig. 5. Nubecula Major (The Greater Magellanic Cloud).

A STAR CLOUD—THE PROBABLE END OF A SPIRAL NEBULA.

[*face p. 77*

identity of these particular spots of light beyond all reasonable doubt.

In these nebulæ, then, we are watching the birth of stars, the transformation of an inchoate mass of gas into an "island universe" of stars. Indeed Dr. Hubble found it necessary to end up his classification of nebulæ with clouds of stars. At one end of his continuous sequence is a nebula, shaped like a mass of rotating gas, in which not a single star is visible; at the other end a star-cloud such as the Greater Magellanic Cloud shewn in Plate V. in which nothing but stars are visible. Our galactic system of stars is probably the final product of just such a transformation, the Milky Way still recording the position of the equatorial plane of the original nebula.

We can form fairly precise estimates of the time since the stars were born, from the circumstance that they were probably far more massive at birth than they now are. If the sun continued to lose mass at precisely his present rate of 360,000 million tons a day, he would last for just about 15 million million years. Similarly if he had always radiated at his present rate, he would have had double his present mass 15×10^{12} years ago, treble his present mass 30×10^{12} years ago and so on. But we know that the rate of radiation is nothing like uniform; observation shews that the more massive stars burn up their substance far more rapidly than the less massive stars, and we may assume that when the sun was many times as massive as he now is, his rate of decay was correspondingly greater than now. Allowing for this we find that the sun would have had

double its present mass		$5{\cdot}7 \times 10^{12}$	years	ago
four times	,,	$7{\cdot}1 \times 10^{12}$	,,	,,
eight	,,	$7{\cdot}4 \times 10^{12}$	,,	,,
sixteen	,,	$7{\cdot}5 \times 10^{12}$	,,	,,
thirty-two	,,	$7{\cdot}6 \times 10^{12}$	,,	,,

We notice that when, if ever, it was thirty-two times as massive as now, it would scatter half of its mass away in radiation in only 100,000 million years, so prodigal are the stars of their substance in their early years when they have plenty to spend. As a consequence of their running through the early stages of the careers at such a high speed, we can estimate the age of the sun, or of any other star, with fair accuracy without knowing what mass it started with. The above table suggests that the sun's age must almost certainly be between 5 and 8 million million years, and is in all probability between 7 and 8.

The calculation involves only the present mass of the sun, and the probable ages of other stars can be calculated in a similar way from their present masses. Stars are like children in that their weights give a fairly good indication of their ages, although no doubt a good deal must be allowed for individual peculiarities. Now when we calculate the ages of a typical sample of stars, as for instance those within a certain specified distance of the sun, we do not find that all ages are equally represented, but there is a quite marked preponderance of stars of ages between 5 and 8 million million years. A large number of stars, then, appear to be of about the age of our sun, and there is a deficiency, although by no means a complete absence, of stars which are very much older, or very much younger, than our sun. Although the situation is by no means free from difficulty, it seems reasonable to suppose that a large proportion of the stars in the neighbourhood of the sun were born by the break-up of a spiral nebula some 5 to 8 million million years ago.

Other groups of stars might of course shew different ages. And we cannot say how long the stellar matter may have been in the nebular state before it formed stars. Nothing in astronomy fixes with any precision the time since

> "the great morning of the world,
> when first
> God dawned on chaos."

Indeed we cannot say whether the

matter of the universe came into being all at once, by a distinct series of creations, or continuously. The general age of the universe is, however, likely to be comparable with that of the sample of stars just discussed.

The Birth of Planets

After their birth stars do not live entirely uneventful lives. They may meet with a variety of accidents and these result in different observed astronomical formations. A star may rotate too fast for safety, just as a flywheel may; when this happens it breaks into two, and the two stars so formed revolve endlessly about one another as a binary system. Two stars may run into one another, although this is very rare. A more common occurrence is for two stars to escape running into one another by a narrow shave. When this happens, huge tides are raised on the two stars involved, and these may take the form of long streamers of gas, which ultimately condense into 'drops' just as did the gas in the outlying regions of the spiral nebulæ. It seems reasonably certain that the planets were formed in this way. The two nebulæ shewn in Plate VI suggest the general nature of the process, although whatever is taking place here is on a stupendous scale.

The birth of the solar system, then, resulted from the close approach of two stars; if a second star had not happened to come close to our sun, there would have been no solar system. It may be thought that with a life of millions of millions of years behind it, one star or another would have been certain to come near enough at some time to tear planets out of the body of our sun. Calculation shows the reverse; even after their long lives of millions of millions of years, only about one star in 100,000 can be surrounded by planets born in this way. A quite unusual accident is necessary to produce planets, and our sun with its

PLATE VI

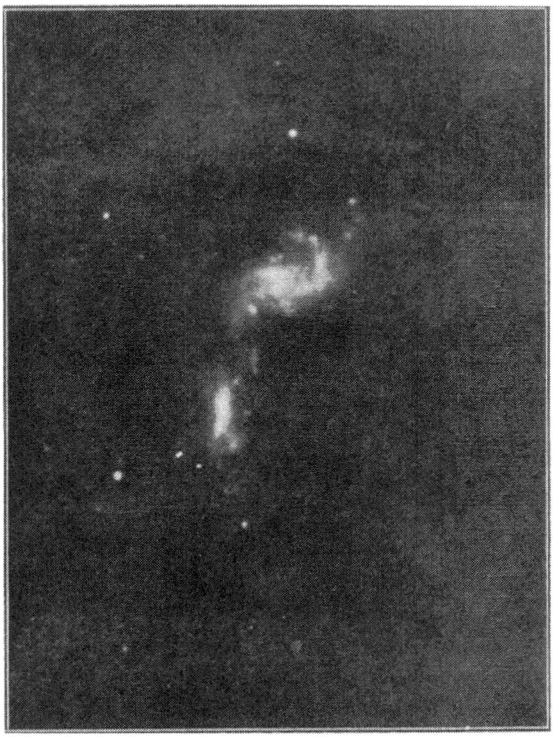

[*Mt. Wilson Observatory*

Fig. 6. Twin nebulae (4395.4401) acting tidally upon one another.

NEBULAE SUGGESTING THE WAY IN WHICH THE PLANETS MAY HAVE BEEN BORN OUT OF SUN BY THE TIDAL ACTION OF A PASSING STAR.

[*face p.* **82**

family of attendant planets is rather of the nature of an astronomical freak.

In the thousand million stars surrounding our sun there are, at a moderate computation, not more than ten thousand planetary systems, because there has not been time for more than this number to be born. They are of course still coming into existence; calculation suggests a birth-rate of about one per thousand million years. Thus we should have to visit thousands of millions of stars before finding a planetary system of as recent creation as our own, and, even if life similar to our own exists on other planets, we should have to visit millions of millions of stars before finding a planet on which civilization, and interest in the outer universe, were as recent a growth as are our own. Utterly inexperienced beings, we are standing at the first flush of the dawn of civilization. Each instant the vision before us changes as the rosy-fingered goddess paints a new and ever more wonderful picture in the sky, while on earth the rolling back of the morning mists discloses new, mysterious and unsuspected vistas to our bewildered gaze. We call it living in an age of progress.

In time the glory of the morning must fade into the light of common day, and this in some far distant age will give place to evening twilight presaging the final eternal night. But we children of the dawn need give but little thought to the far-off sunset.

It may be suggested that the creation of planetary systems is also only at its beginning, and that in time every star will be surrounded, like our sun, by a family of planets. But no; the stars will have dissolved into radiation or disappeared into darkness before there is time for this to happen. So far as we can judge, our part of the universe has lived the more eventful part of its life already; what we are witnessing is less the rising of the curtain before the play than the burning out of candle-ends on an empty stage on which the drama is already over. There is not time for many more planets to be born.

LIFE AND THE UNIVERSE

The planets are the only places we know where life can exist. The stars are too hot; even their atoms are broken up by the intense heat. Nebulæ are in every way unsuitable; even if cool solid bodies exist in them, they would probably be so drenched with highly penetrating radiation as to render life impossible. We have already noticed how life demands a special type of matter, such as does not produce intense light and heat by transforming itself into radiation. We find it only in the surfaces of the stars, which are too hot for life, and in the planets which have been pulled out of these surfaces.

On any scheme of cosmogony, life must be limited to an exceedingly small corner of the universe. To our baby's wonderings whether other cradles and other babies exist, the answer appears to be that there can at best be very few cradles, and there is no conceivable means of knowing whether they are tenanted by babies or not. We look out and see a universe consisting primarily of matter which is transforming itself into radiation, and producing so much heat, light, and highly penetrating radiation as to make life impossible. In rare instances, special accidents may produce bodies such as our earth, formed of a special cool ash which no longer produces radiation, and here life may be possible. But it does not at present look as though Nature had designed the universe primarily for life; the normal star and the normal nebula have nothing to do with life except making it impossible. Life is the end of a chain of by-products; it seems to be the accident, and torrential deluges of life-destroying radiation the essential.

There is a temptation to base wide-reaching inferences on the fact that the universe as a whole is apparently antagonistic to life. Other quite different inferences might be based on the fact of our earth being singularly well-adapted to life. We shall, I think, do well to avoid both. Each oak in a forest produces many thousands of acorns, of which only one succeeds in germinating and becoming an oak. The successful acorn, contemplating myriads of acorns lying crushed, rotten, or dead on the ground, might argue that the forest must be inimical to the growth of oaks, or might reason that nothing but the intervention of a special providence could account for its own success in the face of so many failures. We must beware of both types of hasty inference.

In any case, our three-days-old infant cannot be very confident of any interpretation it puts on a universe which it only discovered a minute or two ago. We have said it has seventy years of life before it, but in truth its expectation of life would seem to be nearer to 70,000 years. It may be puzzled, distressed, and often irritated at the apparent meaninglessness and incomprehensibility of the world to which it has suddenly wakened up. But it is still very young; it might travel half the world over before finding another baby as young and inexperienced as itself. It has before it time enough and to spare in which it may understand everything. Sooner or later the pieces of the puzzle must begin to fit together, although it may reasonably be doubted whether the whole picture can ever be comprehensible to one small, and apparently quite insignificant, part of the picture. And ever the old question obtrudes itself as to whether the infant has any means of knowing that it is not dreaming all the time. The picture it sees may be merely a creation of its own mind, in which nothing really exists except itself; the universe which we study with such care may be a dream, and we brain-cells in the mind of the dreamer.

HERMES

OR

THE FUTURE OF CHEMISTRY

BY

T. W. JONES, B.Sc., F.C.S.

LONDON:
KEGAN PAUL, TRENCH, TRUBNER & CO., LTD.
NEW YORK: E. P. DUTTON & CO.
1928

Printed in Guernsey, C.I., (British Isles) by the Star and Gazette Co. Ltd.

HERMES

Jâbir ibn Hayyan, that legendary father of Chemistry, whose name has come down the centuries in the Western style of Geber, showed his master, Ja'far al-Sadiq, his book, *Kitab al Wasiyya*—the book of his thoughts, his dreams and discoveries.

Whereupon—these are now the words of Jâbir himself—" My master Ja'far, on whom be peace ! was angry when I showed him this book and said : ' O Jâbir, you have revealed the Mighty Secret of God.' (I replied) : ' It was in my mind to be liberal and I wished to be bountiful and generous, for in thy service generosity and truth and frank explanation have remained with me and I have imitated thy nature. But, if thou so orderest, I will burn the book and will deliver it to no one.' He smiled, being well pleased with my words (and said) :

'Do not so, for the Most High God has aided thee in this matter and has made it easy for thee. So oppose not the will of God in the revelation of this book, for no one in our time, or after us, will attain thereto except the Most High God help him and rightly direct him.' "[1]

Herein may be found the statement of faith of the Aristotle of Chemistry: the faith that made him legend and us the inheritors of Chemic Science.

The curious are referred to the authorities Berthelot, Al-Jildaki, Kopp, Ruska, Holmyard, Brockelmann and many another, for information and may find for themselves the truth of the Jâbirian legend and his service of Ja'far, who was the sixth Imam of Medina, in the eighth century, and of his sonship to that politically scheming druggist whose head fell when dread authority discovered him.

It is with the future developments of the Hermetic Art pursued along the lines of exact experiment introduced by Geber

[1] E. J. Holmyard's translation.

and mythologically patronized by Hermes Trismegistus that we are here concerned.

For many generations, unseen and unrequited, the chemist has played his part in the lives of all of us. Recent decades are notable in that his work is now respected, while his economic position is secure. In these later days natural wealth has been relentlessly exploited and waste reduced to its minimum.

The advent of chemical research has wrought violent changes in the nature of some of the world's markets: staple products have been replaced by products first discovered in the laboratory; with these new vistas have come new methods and life itself has changed. Natural and nitrogenous fertilizers, for example, have met, in synthetic indigo and ammonium nitrate, a competition that threatens abandonment of the natural products, and promises a greater command of colour and a fuller fertilization of the soil. Natural products are being quickly superseded by synthetic substitutes at lower cost, in unlimited supply and more useful form.

Chemical science, because it can be applied to the control of nature, has become a necessity to a civilization that daily acquires fresh and more complex material needs. Because, too, this science has now reached a development where it can keep step with these growing needs its importance in the daily life of the community will increase.

There are some, notably Mr. Bertrand Russell in his *Icarus*, who think that the effect of scientific exploitation has been to organize modern industrialism on the narrowest nationalist basis, harnessing patriotic feeling, inflaming greed and the competitive spirit : that the development of science threatens complete overthrow of civilization. They discount altogether the possibility of any economic internationalism except by the coercive domination of a single group.

It is unlikely, however, that the balance

of civilization as we know it can be violently upset. It may quite possibly be destroyed in certain areas, but stability will tend to be restored through the influence of areas not so disrupted. There is also the improbability of any advance in methods of scientific warfare (especially through chemistry) that will lead to more than localized destruction.

Nor do Mr. Russell and others appreciate sufficiently the recent tendency of industrial organization towards international compromise. It is becoming more and more obvious that nationalist monopoly cannot hold any field for long. Increasingly capitalist enterprises are discovering nationalist isolation to be a threat to their economic existence. Thus the liaison between the great firms here and those on the Continent and in America in exchange of personnel, technical information and even financial support becomes closer year by year.

Undoubtedly there will be more wars, interferences and hindrances to this process of economic internationalization, but that it will continue, although slowly, is sure ; and this emerging consciousness of economic unity will supply the stimulus to that moral evolution which Mr. Russell so desiderates.

Science is not, however, to be regarded merely as an economic tool. It is much more. It is the systematized observation and analysis of differing kinds of matter and in its widest synthesis a philosophy of matter.

Necessarily its conclusions must always be empirical. The philosophical position of science as a whole has been revolutionized by the relativity concept. While the biologist and geologist have always realized the dependence of their sciences upon chemistry, the chemist and the physicist, much more the mathematician, recognized no such interdependence ; each regarded his particular branch as self-contained and unaffected by external scientific development. Now our outlook has broadened. Science is no longer envisaged as a series of water-tight compartments labelled chemistry, physics, mathematics, and so on, but as a diversely functioning unity in which any one function is only to be expressed in terms of all the others.

The work of Planck, Einstein, and others upon the mathematical expression of physics has affected chemistry so deeply as to necessitate an entirely new philosophic view-point and adjustment is still going on.

Before the present century all matter was regarded by chemists as rigid, and resolvable into about eighty unchangeable elements as absolute as time then appeared. Now in the light of the New Physics, it appears that matter is a kinetic equilibrium of electrical charges; it has no rigidity; the atoms divide; and the elements, of which there are ninety-two, transmute themselves spontaneously; time is relative; light has mass and the only straight path in Nature can be curved by gravity.

The chemical system underlying matter

is now viewed as one of equilibrium maintained by controlling conditions, change in which shifts the equilibrium either in the direction of increased simplicity, accompanied by partial disintegration, or else in the direction of an increased complexity, usually entailing synthesis into a new substance. The controlling conditions may usefully be described as the degrees of freedom of the equilibrium.

The dynamic atom of the New Physics is essentially such an equilibrium. The atom consists of a central nucleus of positive electrical energy, the proton, around which, in definite orbits, rotate at enormous velocities a number of negative electrical charges, the electrons. This ultra-miscroscopic planetary system *in petto* is stabilized by internal physical forces similar to those balancing the solar system. Upset of the balance appears only possible by the action of external agencies such as collision with fast-moving electrical particles, and irradiation by light of extremely short wave-length. The atom (except of radio-active elements, which undergo disintegration under normal conditions), appears to require for its disruption external forces so violent that Dalton's concept of it as ultimate and indivisible is understandable. Understandable also is the success attendant upon the efforts of Chemistry during the last century to build upon it theories of its chemical relationship that worked so well in actual practice that Professor Armstrong could say of organic chemistry: "We are able to bet in terms of several hundreds or even thousands to one, having wagered and won with such uniform success during a period of over fifty years."

The sensationalist press and the novelist of the fantastic have led the public to expect early and far reaching developments through the application of what is vaguely termed "atomic energy." But the cost of artificially disrupting so stable an electrical system as the atom does not bring it within the range of practical usefulness. Experiments show that on the average the cost is something like 100,000 times the gain, and there is no evidence that the cost could be decreased to an economic figure. It has been calculated that 1 gramme of radium would have to project its rays into a sheet of aluminium for a period of 3,000 years to disrupt the aluminium atoms sufficiently to yield 1 cubic millimetre of hydrogen. Scarcely an explosive reaction !

Under normal terrestrial conditions the atom may be regarded (for general chemical purposes) as a stable entity but (and this was where the Daltonian theory failed), the electronic configuration, the position of the electrons around the nucleus, is always vital, always germane, to every chemical inquiry.

This electronic configuration, so far as we know, is the sole predisposing cause of chemical and physical combination of the atoms. The manner of the internal arrangements of the electrons confers a condition of unsaturation upon the atom which it seeks to satisfy by combination with other atoms to form molecules.

Although the peculiar behaviour of every atom and every molecule is traceable to its electronic equilibrium, yet that behaviour is its chemical nature, and is the cause of chemical change. Because the equilibrium of every molecule is comparatively unstable chemical change can be arbitrarily controlled by man: he can vary the degrees of freedom of every molecule at will.

By altering the degree of freedom of temperature, for example, he can induce substances unattracted to each other at ordinary temperatures, to combine. Carbon and oxygen can exist side by side in perfect tranquillity, heat them together and they combine to form carbonic acid gas. Other substances, on heating, undergo disintegration; for instance, chalk, at elevated temperatures, breaks up into lime and carbon dioxide.

There are other degrees of freedom beside temperature; one is pressure, another is surface tension and another actinic light. Of these surface tension has yielded extraordinarily important discoveries: at least, we think the phenomena observed are ascribable to surface tension. If two substances are brought together in the presence of a third, this third body brings about combination of the other two substances. This *tertium quid* is called the catalyst, and probably performs its function by condensing the atoms or molecules, of the two substances on its surface, in layers one molecule thick. In this condition they are in a state of surface tension, the electronic configuration of both of them is then highly activated and they are in unstable equilibria and may combine to form a new molecule. Whilst apparently every chemical change may be catalytically induced, the catalyst acts selectively; it is not suited to promote the combination of many different compounds; in fact, it is usually found that for every chemical reaction there is but one wholly suitable catalyst. For example, the catalyst best suited for the hardening of vegetable and fish oils, by combining them with hydrogen, is nickel. Nickel favours this particular chemical change most when in the finely divided form known as the colloidal condition of matter. It appears to be the most necessary requirement for catalysis that the catalyst should be in the colloidal state. In the colloidal catalyst chemistry reaches up to, and imitates, living matter.

In organic chemistry there is a dependence on structure which masks the electronic equilibria of the associated atoms. In fact, so dominant is this structural arrangement of the carbon atoms that their chemistry is the most homogeneous branch of the science. Predictions of the behaviour of any newly synthesized molecule can be made from a study of its structure alone, with the accuracy that Professor Armstrong's words predicate. This property of structural integration separates the carbon atom from the rest of the elements, except possibly boron and nitrogen; in all probability, although we cannot say so with certainty, it is the property from which arises its employment in the structure, without exception, of all living matter.

The molecule of living matter, like any other, exists in a state of equilibrium changed only by alterations in external conditions, the degrees of freedom; by reason of its internal structural arrangement, it possesses to a marked degree a certain adaptability to external conditions. This indicates an internal adaptability, although of only an elementary degree, to environment—a property peculiar to the living organism.

The rôle of the catalytic degree of freedom becomes more important and more selective the further we advance up the range of organic molecular complexity until, in living integrations of matter, it reaches its apogee. It is natural that this should be so, for the living organism depends for life on the change of its food into molecules which it can temporarily synthesize into tissue and then break down again. Temperature and pressure are degrees of freedom that cannot be used by the organism, since it has only a limited tolerance for them. The organism has, therefore, to rely on substances which work independently of high temperatures and pressures to convert its food into simple molecules and then to synthesize them again into tissue. These substances, which most probably act catalytically are exceedingly complex, and appear to possess a peculiar adaptability which enables them both to break up, or hydrolyze, the food molecules into something simpler, and then synthesize the resultant molecules again into complexities required by the organism. Hence, in living matter, the catalysts by which life is maintained are kept in a condition of equilibrium so that the sum of their synthetic and hydrolytic catalyses may be balanced in the direction required for the furtherance of the living whole of which they form a part. Since disintegration is death, the balance of the vital catalyses must be weighted on the side of synthesis, or, as Jacques Loeb put it: "The matter between life and death is not between living and dead molecules, but more likely between excess of synthetic over hydrolytic processes." "La vie, c'est la création," said Claude Bernard.

So far as we have investigated them, the vitamins appear to be the requisite external degrees of freedom to control this equilibrium in human life; they balance it in favour of the synthetic process, for in their absence the so-called deficiency diseases supervene, and lead, finally, to death. They appear to act catalytically to balance the syntheses of life over the hydrolyses of death. One may perhaps regard them without incongruity as the super-catalysts of vitality controlling the body's own catalysts. Unfortunately, we know as yet very little about them, but, as far as we are in a position to judge, the human body seems unable to make the vitamins it requires or to store them for very long ; it requires constant supplies of them ; yet the lower forms of life, the micro-organisms and plants appear to make them and to store them with ease. Man and the higher animals, it seems, depend entirely upon these lower forms for their supplies. Whatever they may be, and however they may work, the vitamins appear to act as most potent degrees of freedom for the animal equilibrium.

The chemist then finds that all matter consists of individuals, or wholes, as General Smuts in his *Holism and Evolution* has called them, which are each and singly maintained in a stable condition of equilibrium by degrees of freedom, external and peculiar to the stabilized system. As matter is integrated, so are its degrees of freedom ; for every whole and individual integration there exists its peculiar degrees of freedom which alone maintain its inner harmony. Yet, in all matter, and especially so in living matter, there appears to be an internal selectivity that utilizes, within the limits of its equilibrial range, the degrees of freedom for its own purposes.

This concept of the self-regulation of the whole is, of course, implicit in all Loeb's work, but both he and Smuts appear to miss the significance of the degree of freedom. This self-regulation, as integration proceeds, appears to become more and more selective until, when we come to life, it may possibly be synonymous with life. If we do not find it so, then there appears no alternative to a mechanistic philosophy wherein the degree of freedom becomes identified with the external directive force, so at variance with all Loeb's ideas, that was put forward by Claude Bernard.

It is extraordinarily significant of the homogeneity of matter that its chemical nature, in whatever degree of integration examined, is found to be based on one single system the balance of which is revealed as dependent ultimately when traced to the foundations, upon the electronic system of the atom. It is evident that all science must be as homogeneous—as consistent—as the matter it studies : the sciences of the Victorians are all one, and chemistry but a phase of scientific discovery and progress. But at the same time sight should not be lost of the fact discovered by chemistry that for all practical purposes of living matter the electronic condition of the atom is only displayed chemically ; hence the study of the higher forms of matter rests on a chemical foundation, to which they may be reduced, and upon which they are built. The application of chemical science to matter is, therefore, the only efficient method of controlling it for the needs of mankind. As yet man is scarcely aware of this fact, but he is learning fast and moving with sure steps to the chemical exploitation of material resources ; a necessity to the overgrowing complexity of life.

As a dynamic entity man needs energy, and he obtains it from his physiological use of food. In the civilized state of an organized society, however, wherein the conditions of urban and similarly densely populated areas necessitate the wholesale transportation of food, he must have an additional form of energy, namely, fuel for locomotive purposes. To Isolated Man, as the economists have christened him, food is the essential and mechanicalized transport a superfluity. What direct service unobtainable from his own hands and feet could Robinson Crusoe have got from an aeroplane ? Only that of escape, which, once achieved, would have immediately disqualified him as a self-contained economic unit.

To commercialized man, on the other hand, transport must be ranked as one of the primary needs, if not the first essential, of his civilized state. Without it he must face reversion to a condition of barbarism approaching more or less nearly to that of the economists' Isolation. The highest state of social integration possible under the conditions arising from an entire elimination of every form of transportation more mechanical than the horse and cart would be akin to that of the eighteenth century when famine was not unknown and was, indeed, a possibility to be reckoned with by most of the population. The last state would probably be worse and dearth become the seasonal event it is in all primitive societies. It would very likely be possible to show that lack of mechanical transport, as well as disease, lay at the root of the troubles of all earlier civilizations and without it they were doomed to disintegration. Since we may expect still greater improvements in transport we may assume progressively higher integration, as transport is the keystone of social life and culture—food is but that of population.

The transport of the future, one may suppose, will be mobilized by electrically and thermally driven engines, whether it be road or air traffic.

Since our supplies of coal and petrol are limited to the amounts stored in the earth, the day will come when they are exhausted; but new fuels will have been synthesized to take their place as sources of power for thermally-driven traction, for it is probable that for certain types of transport we shall always rely upon thermal power ; fuel is easy to carry and store, and confers on the vehicle using it independence of movement, complete mobility ; whereas the mobility of electrically-driven vehicles is restricted either by the conductors through which they receive their power, or by the capacity of their storage batteries ; the superior mobility of the thermally driven vehicle makes it a factor of cardinal importance, although our harnessing of the water power of the world to hydro-electric turbines could probably supply all our demands for energy for every purpose, including that of transport. Indeed some of the hydraulic power of the future may be diverted to the electrochemical synthesis of fuel ; we may find some mineral other than coal to supply the requisite carbon to burn in our combustion engines.

To return to " atomic energy," briefly

touched upon earlier. It has long been prophesied that in the future we shall supply all our needs of energy directly from that of the atom ; the energy of the atom is said to be such that a cupful of water could drive the finest of the Cunard fleet from Europe to America and back. It would probably take the hydro-electric power of Lake Erie, spread over a number of years, to disintegrate completely the cupful of water. Furthermore, the prophecy argues from false premises ; it takes for granted that the nature of atomic energy is known and differs from other forms of energy ; whereas we can, at present, only guess at its exact nature. We know that its equilibrium is electrical. We already use electrical energy, and, if we are to gain nothing more than we have at present, it is difficult to see wherein lies the advantage of this over-rated "atomic energy." The most we can say at present is that it *may* turn out to be a fresh material force, and *may* be harnessed ; but there is not, at present, the slightest indication that it will, nor that our traffic problems of the future will be resolvable by any energy other than that derived from our resources of water and fuel.

Now it is only a matter of time before our petrol reserves are all used, and we have to find some other suitable fuel in equally enormous quantities; for it is not to be expected that the benefits of the light internal combustion engine, thermally a highly-efficient apparatus, will be surrendered either by business transport, or by *paterfamilias*, who, over the week-end, derives from it much of what the economists assure us is an indirect satisfaction of economic need. There is little doubt that we shall find this fuel, and find it in the chemical factory; probably long before our supplies of petrol are exhausted. Once it can be produced in commercial quantities it will be in strenuous competition with petrol. The decisive factor in the disappearance of petrol from the world's markets may then well be the strength of counter competition it can achieve, rather than the exhaustion of its deposits. Mother Earth could then keep her stores of mineral oil; it would not pay man to take them. It was at one time predicted, and no longer ago than 1905, that the deposits of Chilean nitrates would be exhausted and the world left without fertilizers for its crops: to-day Chilean nitrate meets only a third of the world's needs, and except for a fraction made in gas works, the remainder is made synthetically. Perhaps petrol will follow this example and oil-profits dwindle as the nitrate profits have done; it all depends on the length of life of the petroleum deposits, and the cost of its production. Synthetic substitutes are imminent.

The eccentricity which Nature, like a French peasant, displays in hiding her wealth, renders all man's computation futile. Estimates of both must be limited within predictable bounds, but although the peasant's maximum possibilities may be judged with fair accuracy, by a careful observer of the habits of his class and race, nobody has the requisite knowledge to define the extent to which Nature may have saved. Estimates vary as widely as eight to eight hundred years. The best informed place the oil left in the wells of the United States, the biggest single producer of oil, at barely six years' supply, with a problematic twenty years' supply from shale deposits. True, the United States is a single producer, but in 1924 those wells yielded 70.5% of the world's petroleum. In the same year American motorists consumed 73.8% of the world's petroleum. The most prolific producer found his supplies insufficient!

The near future will witness the manufacture of a number of fuels. Some will be heavier grades of oil, yet still competing with petroleum fractions for the patronage of owners of Diesel and similar heavy engines.

At the present time there is extraordinary activity in laboratories all over the world to this end: much of it will be abortive—man has to experiment to learn

—but the net result will provide an extremely wide range of fuels suitable for all kinds of engines. It would be premature to enumerate them: they all rely upon obtaining from coal of some kind, the cheaper the better, an hydrogenous liquid product possessing properties as closely resembling petrol as possible. The methods of getting them vary; some compress the coal with hydrogen and a catalyst, as in the process known as the Bergius process, which yields a liquid very similar indeed to petroleum: some use coke which they burn partially to carbon monoxide, combining this with hydrogen catalytically at high temperatures and pressures to obtain a simple alcohol, methyl alcohol. This particular process is now being carried out on a big scale in factories both in Europe and America; however, it is not the best of fuels for the ordinary motor car engine; but the process is interesting in that, by it, under slightly different conditions of temperature and pressure, it is possible to obtain certain compounds chemically identical with those in ordinary petrol. Another method, that is being worked on a large scale in Germany, heats together coke and lime, from which by way of acetylene, ethyl alcohol, ethanol, is synthesized catalytically. So far none of these is competing with petroleum, but when they are developed a little further and processes cheapened sufficiently, they will come into the market for that purpose, besides, of course, those they now supply. Their advent is only a matter of time ; probably by 1935 other synthetic fuels will have been evolved by the chemist, and the motorists of that date will find a difficulty in making a choice from the numerous pumps then lining the kerbside.

As all these synthetic fuels are derived from coal the duration of their utility must depend upon the life of our coal supplies. Here there is no cause for apprehension of a shortage in the near future. The world's supplies of coal are comparatively enormous, being estimated at about a hundred billion tons; and it is as well they are, for they are annually drained at the rate of over twelve hundred million tons; a tonnage approached by no other commodity. Since the average yield from a ton of coal is fifteen hundredweight of coke, practically all of which may be used for synthetic purposes, it is obvious that a final extinction of the motor-car through lack of fuel need not be feared. Because a growing use of transport of all kinds coupled with a probable exhaustion of petroleum, or at all events a diminished production of petroleum in the near future will make increasing inroads upon the coal reserves, the present estimated life of 3,000 years for the coal-mining industry may have to be reduced. But it appears that, in all probability, besides those economies resulting from the hydro-electric developments of the future, others will arise through changes in the use of coal.

At the present time most of the coal is burned in the same condition as it is mined; volatile constitutes, so valuable for the manufacture of synthetic dyes, drugs, and perfumes, are wastefully sent up the chimney as smoke ; befouling our cities with an atmosphere that effectively clogs them against penetration by sunshine, and prevents the entry of fresh air. A change in our use of coal would have in its support the forces of cleanliness, health and economy. These, however, do not appear sufficient to persuade us of our errors, and the Royal Commission of the Coal Mining Industry, in 1925, actually hinted that it might be advisable to add the force of legal penalization in order to ensure adequate pre-treatment of all coal.

Of the various methods suggested for pre-treatment, that of distillation at low temperatures is the most generally supported at the present time. This is claimed to be the most efficient, as it extracts from coal the majority of the volatile constituents, including gas and motor fuel (the yield of benzol—the motor fuel directly distilled from coal—is three times greater than that obtained by the usual gas works practice), a tar containing all the essentials for the dye-stuffs and disinfectant-making industries, and a smokeless coke that can be burned in domestic grates as easily as raw coal. Incidentally recent experiment has shown that the open fire of the English domestic grate is healthier for the occupants of the room than any other type of heating—we may yet live to see America and the Continent copying our so-called antediluvian ways of heating.

This method of low-temperature distillation almost certainly will become general practice when various technical difficulties have been overcome; for, despite the claims of its enthusiastic supporters, who point to numerous experiments, in Glasgow and elsewhere, there are still drawbacks that, at the moment, make it commercially unprofitable. It is, however, the nearest approach we have yet made to the ideal use of our most valuable natural heritage; the nearest approach to the mode of exploitation of mineral wealth an economically determined community might be expected to adopt. At present the only economically-minded members of the community appear to be the owners of gas-works, and they, it is claimed, waste more than they would if they worked their coal at a lower temperature. It is not, however, on the increased yield of motor spirit that the supporters of this method base their advocacy—indeed, it is difficult to see how they could, since, were *all* the coal annually mined by the world so treated less than a tenth of our present requirements for motor fuel could be met—rather do they rely upon its promise of the abatement of smoke ; in consequence of the artificial removal of all the volatile constituents of the original coal, the coke burns freely and without smoking ; consequently, there is no smoke to befoul the atmosphere, to dirty buildings and encourage the spread of phthisis among their inhabitants. The supporters believe it will ensure cleanly cities and healthy citizens. Whether governmental coercion, as hinted in the Royal Commission's *Report*, will be required for its universal adoption is doubtful, it is more likely to result from private enterprise by reason of the control that could thereby be exercised over the raw material of both the chemical and synthetic fuel industries of the future.

The pronounced tendency on the part of modern manufacturers towards vertical combination, to ensure adequate supplies of raw materials, leads one to expect that the coal supplies of the future will be controlled by chemical manufacturing syndicates. If this should come about, it may well happen that a large proportion of the world's coal will be subjected to preliminary carbonization for the sake of its chemically valuable distillates; such of the coke produced as may be required for synthetic fuels will be retained, and the remainder passed on to the domestic and industrial consumer. The householder of the future may find that the only fuel available to heat the home will be smokeless, and he will be prevented, willy-nilly, from polluting the atmosphere. Unlike his modern prototype, who is responsible for most of the dirt of the urban atmosphere, he will enjoy in a town dwelling all the freshness and sunlight of a country climate—gas-works will be relegated to the mining districts, whence all gas will be distributed over the country by pipe-lines.

Further economies in the consumption of coal are bound to come by the development of hydro-electric schemes. The electrical power will, by the ingenuity of the electrical engineer, be distributed to all parts of the earth's surface, gradually supplanting thermal power for all purposes except those of independently mobile vehicles: the numbers of these will finally be reduced to the minimum. Railways will probably be the first form of communication to be completely electrified. A close second will be the bus-routes; the electric tram may at last vindicate the faith of Manchester, and the motor bus may then be superseded both for urban and country services by some form of the trackless trolley.

To-day the chemist's most valued assistant is the electrical engineer; with his help he first made synthetic fertilizers from the air, and separated aluminium and beryllium from their ores.

Judging from the proceedings of world power conferences the activities of the electrical engineers are nowadays devoted to hastening the final conditions of terrestrial configuration wherein every valley shall hold its lake, the annual rainfall for which is assured by forests—spreading from the water's edge to the snowline—in which the holiday-makers of the future wander during summer days; and in the winter the proprietor or, probably more accurately, the holding syndicate, cuts the areas scheduled annually for the making of paper and other commodities future discovery may make possible from timber, bark and roots (metal will have superseded timber for building purposes). At the lower end of the lake small boys slide down the pipes to the turbine house. In that millennium all houses will be electrically heated and lit, cooking will be electrical, as we shall at last have exhausted our coal and be dependent upon the carbonates of the earth for carbon. Chalk will provide the fuels of our aeroplanes and motor cars. Chemical science will have evolved an electrical conductor that will efficiently transmit current from Swiss turbines to the cliffs of Dover, to separate pure carbon from the chalk. Electrically driven pumps will deliver the waters of the Straits to electrolytic cells on the cliff-top from which the hydrogen will pass to a catalytic chamber and there, with the carbon, be synthesized to a fuel better than any we have ever known. Visiting Gauls will look in vain for the historic landfall of Albion.

It is difficult to foresee any innovation in the principles of vehicular construction whether designed for aerial or terrestrial service. We seem to have exhausted the possibilities. Vehicles will probably continue to be boxes of one kind or another, open or closed, provided with skids, wheels, keels or wings. New methods of moving them are improbable ; we have driven them by external agencies—animals, wind and gravity, and by internal engines—steam, electricity, gas, and man-power.

We may, and probably shall, see changes in constructional materials. Even here the novelty will be confined to degree rather than kind. The constructional material of the future will be a metal, or rather alloy of metals, mainly for the reason that our timber will have been exhausted ; stone and cement being dismissed as obviously unsuitable. The supply of metals may be taken as inexhaustible ; the whole of the earth's surface, so far as we have penetrated, consists of chemical compounds of the metals, the only obstacle to their use is the difficulty of separating them. This will be surmounted in time.

The need of aeroplane construction is

obviously one of lightness combined with strength. For this we must rely upon the lighter metals. But they have the disadvantage of mechanical weakness, unless suitably alloyed. Promise appears to lie in the direction of alloying aluminium and beryllium, very nearly the two lightest metals, and research upon these is being pursued diligently both in Europe and America. A German firm, the Chemische Fabrik Greisheim Elektron, claims to have produced a suitable alloy somewhat less than twice as heavy as water. Safety, particularly for trans-oceanic passengers, will be assured when an alloy capable of floating upon water has been invented; lithium would do as an alloying metal, but it would either have to be crystallized in some special way or alloyed with a heavier metal to increase its strength and chemical resistivity. Travellers of the future may demand a chemical analysis of their aeroplanes before trusting themselves in the air between London and New York.

It is long since the roof-tree was the only alternative to the cave. In considering building materials of the future we have to take into account a certain exhaustion of our timber supplies ; an exhaustion, moreover, that according to the experts, will not be long delayed. It is computed that already European demands exceed the annual growth of soft woods by about three thousand million cubic feet. Soft woods, it is true, are not bought entirely by the building trade, but are taken in bulk by the paper trade. But they are quick growing and hence offer greater attractions than the slow-growing hardwoods to the speculative grower, and we may take it that the annual replenishment of our supplies is confined more or less entirely to these soft woods. It would seem, therefore, reasonable to suppose that the annual proportional excess of consumption over growth for building timber is definitely on the wrong side. Supporting evidence is not lacking in the price of the better quality hardwoods like oak, mahogany, maple, and so on, which are, except for what may be termed *de-luxe* building, used exclusively for furniture making. Presently the hard woods of the furniture trade will be replaced by the better-class soft woods, such as pine, although it is doubtful whether the paper industry will not outbid all other buyers, even though wireless telephony and television modify the demands made by the daily press. In that case the furniture trade will have to rely, like the building trade, on metallic materials : fortunately chemists have already made some preparation for the eventuality : it is said that quite pleasing and permanent decorative effects can be obtained on aluminium and copper by suitable chemical and heat treatment.

Steel and iron are the very bones of the modern city ; carefully articulated steelwork is the chief support that permits us to crowd upon a few acres of soil. The multitudes that apparently must be concentrated for the maintenance of commerce and administration, whether it be office or factory, have made inevitable the superposition of one floor on another. The bungalow plan of the cave is useless to business, however suitable it may be for dwelling purposes. We shall find metals used more extensively in dwelling houses, as timber becomes scarcer; we already have metal windows, baths, and fireplaces; we already have, even if experimentally, metal doors, picture-rails and stairways; metal office furniture is to-day selling readily in London.

All-metal houses cannot be expected to become the final word in accomodation that all-metal vehicles obviously are in transport; we have tried them, and must doubt whether the idea will survive the types erected in 1920—1926 which were planned to meet the housing shortage.

The cheapness and the general utility of bricks, stone and cement, make their supersession by metal very improbable, at all events for external use. Cement offers most possibility of development; the builders of the Middle Ages showed us what could be done with stone, and the Georgians the resources of brick ; we shall possibly show the future the plasticity of cement ; it remains for the future to build with synthetic glass and other organic plastics—if the raw material can be found in sufficient quantity.

The first thing we have to do with cement is to invent an impervious lightweight variety of sufficient internal tenacity that it may be poured *in situ* to form thin, continuous roofs, quick-setting so that it may be used for the sloping roofs of northern climates. This will assist considerably in reducing building cost to a level we have almost forgotten. Cement has the advantage over other stones, in that being chemically compounded the ingredients may be varied at will, to yield a variety of results ; to-day chemists are busy exploring its plastic and harmonic potentialities. Perhaps one day the architect will arise who can really build in form and colour with cement, instead of the restricted lines into which it is compressed to-day: lines more suggestive of the rigidity of timber than the grace of a freely-flowing mass.

With regard to the clothing of the future, the clothing manufacturers will undoubtedly compete with the paper industry for its raw materials—or rather the textile fibre makers will—for these at the present time find in sulphite wood pulp their cheapest raw material (cotton, the alternative form of cellulose, to-day costs approximately 5d. per pound as against 2d. for the pulp). If a diminishing supply and increased demand by the paper makers have the effect of raising the price of pulp, obviously cultivation of cotton will be increased, and the day may come when the cotton field is very nearly the sole source of clothing, and possibly also an important source of news print.

The chemist finds in artificial fibres the most fruitful solution of man's growing shortage of clothing. At present, in point of quantity, cotton is the chief textile fibre;

the second is wool; next comes artificial silk, and among the rest is natural silk. After the fig-leaf and skins, wool was man's first covering, and so continued through the Middle Ages up to the beginning of the last century; then, thanks to the slave trade, cotton supplanted it. To-day artificial silk runs next to wool, and in time will probably overtake it, as cotton did. Cotton has succeeded by its cheapness and ease of cultivation: the yield per acre is considerably more than from wool: the crop is annual and seasonal fluctuations are more easily wiped off over a period. Cotton, however, fails by its lack of brilliance and its general lack of artistic qualities: it is a dowdy textile. Treat it chemically and it equals silk. This is the secret of the success of the artificial fibres, which are derived either from sulphite wood pulp or cotton fibre; for little more than the cost of dowdiness they make their wearer the glass of fashion; by chemical art, social barriers are transcended and factory women can closely follow " the mode." Natural silk is a waning industry, all the silkworm areas are now building viscose and cellulose acetate factories; Lyons and Italy are leading the change.

Artificial fibres are superior to cotton in other ways as well; their manufacture is independent of season and situation; employment in the industry is steady with none of the fluctuations and uncertainties of silk weaving; they cost the wearer less whilst benefiting the manufacturer more. Hence their expansion will continue until a more economic solution of the clothing problem is invented.

Mankind's clothing marks the several stages of our civilization. Anatole France and Thomas Carlyle are agreed on this, and we may accept their testimony as valid. The conventions of civilization were roused the instant France's Saint clothed the penguins: " Clothes gave us individuality, distinctions, social polity ": " Clothes made Men of us," wrote Professor Teufelsdrockh. Indeed, through all

Western literature runs the thought that clothes have a civilizing influence.

Given the same population, the same quantity of clothes will be required, no matter the extent of sophistication. Since however, we are to expect an increasing population, we have to seek increasing quantities of fibre to maintain it.

The source of future supplies will undoubtedly be cotton, at any rate, so far as the bulk of clothing is concerned, since supplies of sulphite wood pulp may not more than meet the requirements of the daily press; other annual crops will continue to supply a part, as they do at present; as, for example, linen, hemp, and other bast fibres. But cotton is the purest natural form of the important cellulose molecule and as shown by the statistics of the past century, its rate of growth is sufficiently elastic to meet all demands upon it, and by chemical treatment its natural condition may be altered to suit any purpose for which it may be required. Artificial control, whereby the cellulose molecule can be synthesized into something more complex, places the whole question of clothing entirely, and at last, within the scope of chemistry. No longer shall we remain content to waste energy in attempts to imitate animal furs and wools, and vegetable stalks and seeds, so that what protected the beast or plant from sun and rain shall also protect us. So far manufacturers have confined their efforts to imitations; to out-silk silk and out-wool woolliness; presently they will awake to the fact that the ideal fabric has yet to be evolved—we have contented ourselves with a second-best.

The imitative stage in colour has long been passed. Coal tar dyes, thanks to Perkin and the Germans, are synthesized no more to imitate the colours of nature, whether of autumn or spring; the pattern cards of dye-stuff firms display a multitude of syntheses that transcend Nature to reach almost a philosophic satisfaction of the aesthetic sense of colour. When the problems of durability, both as to washing and sun, and—such is, unfortunately for the Higher Thought—the effect of the body on its clothing—to sweat also have been solved, the corporeal needs will have been satisfied.

Artificial fabrics are synthetic in a small degree so far as they are so by human agency. The major part of the synthesis is carried out by the plant, which can build up complex molecules from the soil and the atmosphere far more cheaply than is possible in the factory. It is only the final stages that are commercially profitable in the laboratory in the case of a molecule as complex as cellulose. Further development of the natural cellulose resources of the earth will lie in assisting the natural synthetic processes, by evolving a plant to yield the maximum of cellulose and providing it with soil containing its food in the simplest condition. Development economically will be in the breeding of plants capable of productivity in climates now unsuitable. We may find it necessary at some time to extend the cotton belt to latitudes where at present cotton is unknown, in the same way that the wheat belt has been shifted in Canada.

A product depending upon cellulose is the nitro-cellulose paint or lacquer. This may compete severely with the textile trade for the cotton which is its raw material—the final product being similarly partially synthetic. For this the highest quality of cotton is required, although doubtless, with improvements in technique less valuable qualities could be used.

Nitro-cellulose lacquers and paints in the very near future may form the sole means of protecting all exposed metal and wood-work ; they will replace the oil and shellac media now used. The nitro-cellulose lacquer or paint is, to quote an authority, " unaffected by sun, rain, snow, or boiling water, salt air or alkali dust " ; claims which cannot be made for other protective coverings ; if they reduce the world's paint bill by only a quarter, they will have justified their invention. It is probable they will do more than this. Indications of their growing use are to be found in the fact that, whereas in 1918 they were unknown outside the chemical laboratory, in 1925 Great Britain and the United States, between them, manufactured 11,075,000 gallons. Paints and varnishes as we know them to-day obviously cannot sustain such competition. Their going will make economic changes of world-wide importance. Of their ingredients, some will find new markets; others, failing this, will cease to be produced, thus compelling workers all over the world to seek more modern occupations. The turpentine production of the world will probably not be changed since turpentine is already wanted for conversion into synthetic camphor for the celluloid and nitro-cellulose lacquer trades. The rosin may be similarly treated or will form a raw material for the synthetic resin trades. The oils will be the most difficult problem. The most used of them is linseed oil, obtained from flax seed. It is thus a by-product of the linen industry and unless it can be absorbed in some manner, the difficulty of disposing of it may result in an increase in the price of linen and, possibly, a gradual dying of the linen industry. The only use the oil could be put to would be the manufacture of foods, since the soap trade would presumably be already flooded with other oils, notably the cotton seed oil from the increased cotton crops. In addition to linseed oil, which is mentioned specifically on account of the enormous quantities produced and the effect its disappearance from the market would have on the linen trade, a use would have to be found for the other oils of the paint trade. Fortunately catalytic hydrogenation would render them valuable to the food trades. In this, however, we are already anticipating the future. At the present time, something like 300,000 tons of vegetable and fish oils are hydrogenated annually by European factories alone, and a large proportion of the resulting fats are used for food.

For many years prophecy has been busy

with the food of the future, predicting such a variety of diets as to make further prediction appear, not so much superfluous, as fantastic. But there seems little reason to suppose that the final food will differ greatly from the present except in regard to its digestibility—in which there is still great scope for improvement. Vegetables and fish will tend to predominate, but butcher's meat, on account of the high economic costs of its production, will probably disappear from the diet of a civilization forced to depend more and more upon the higher yields obtainable from the acre by annual crops.

The foods synthesized from coal and coal tar that have frequently been predicted are never likely to materialize; the cost of their production would make them a luxury in any age and particularly under conditions wherein proper cultivation will yield cheap and plentiful supplies of natural foods, or natural foods slightly elaborated by synthetic art. Partial synthesis, whereby an inedible substance can be made palatable and of dietetic value will always be a commercially profitable practice provided the costs of operation are sufficiently low to enable it to compete with the natural food as, for example, they are in the case of the catalytically hydrogenated vegetable oils used to-day for margarine and other fatty foods. But complete synthesis, from their elements, of complex molecules like the carbohydrates, proteins and fats of foods, it is safe to predict, will be too costly except for special purposes such as laboratory experiment. Simple molecules, like alcohol and ammonium sulphate, involving only a few, comparatively simple, processes, are already profitably synthesized from their elements; but simple molecules are not suitable for the human diet except in small quantities at a time, for man's body is built and run on complex molecules.

The value of food is measured by the amount of energy, estimated as heat, into which it can be converted. This perhaps is not the ideal measure, but it is the best we can do at present, and it served us very well during the war. For a man to do a day's work, repair the usual wastage of the body and put by reserves, the total heat value, according to the well-known Voigt ration, should be 3,000 calories. This is provided by 100 grams of protein, 100 grams of fat, and 400 grams of carbohydrate—carbohydrate, fat, and protein are the three chemically generic classifications of all food. Of the three, the protein, or nitrogenous class can be used for both body-building and supplying energy ; the other two, the fats and carbohydrates, the latter including the starches and sugars, only supply energy, and the fats more slowly than the carbohydrates. Protein and fat have been in the past provided in a diet restricted to meat, hence the ability of primitive and nomadic peoples to live by hunting and their flocks. In addition, the body requires supplies of salts and the newly-discovered accessory food factors, the vitamins.

The vitamins appear essential to healthy, normal life ; without them growth ceases, disease runs riot and the body soon dies. Fortunately for our early ignorance, a normal diet contains all we need of these essential *elixirs* of life. What they are, whether new and unknown molecules or molecules already known, but specially activated, we have not yet discovered ; our only information is that there are five types—there may be more—lack of any one producing the so-called deficiency diseases peculiar to the type.

Considering the complexity and definiteness of these requirements, it is exceedingly difficult to understand how a hearing was ever granted to some of the prophecies of the liberties man has been expected to take with his stomach. Take the tabloid theory, in which a tabloid a day is to satisfy man's hunger and provide him with enough energy for work and play, health and strength, grief and love. One can but imagine that the tabloid, weighing, presumably, a couple of grams at the most, would consist of a selection of extremely efficient catalysts and vitamins, capable of rendering man able to synthesize from the air the remaining 598 grams of food required by the Voigt ration, perhaps as the plants do by the aid of chlorophyll. It is a pity that man's skin cannot do this, as the clothing problem as well as the feeding problem would be solved, since man, one concludes, would go naked in order that his body would deal with the maximum amount of air; incidentally, it would throw out of employment a large proportion of the population. Lancashire and the East may breathe freely: man will continue to require the equivalent of 3,000 calories, woman 2,500, and the baby somewhat less.

Chemical science, already playing an important part in the preparation of natural foodstuffs, is bound to influence, and, to a certain degree, to control the development of both the nature and extent of future food supplies. The foods of to-day, except for the gathering of a few casual crops, like blackberries and mushrooms, cost the populations of the civilized world much more than merely the trouble of harvesting and eating. Throughout there is a highly organized preparation of food for the consumer. In this the chemist bears an increasingly active part, providing disinfectants for the seeds, manures for the ground, other disinfectants for the plant and fruit, supervising packing, treatment and distribution, to ensure that the consumer obtains his food in the best of condition, whether it be wheat, fruit or meat, fish, flesh, or fowl.

In the future, besides this work of a more or less routine nature, chemical activities may be expected to expand very considerably in view of the developments that are foreshadowed by contemporary research in the breeding of plants and animals for food : soil irrigation and, not least, the internal organization of the plant present essentially chemical problems exercising fundamental influence upon both the quality and quantity of food obtainable from them.

The maintenance of numbers as required by growing density of the population appears to be impossible upon the comparatively sparse yield of animal food obtainable per acre. It is significant of modern conditions that in highly civilized communities where in economic phraseology, the marginal returns from land are so narrow as to occupy the minimum of space on the farmer's balance-sheet, intensive growing of crops is the rule rather than the rearing of herds. The production of meat is relegated to the wilder lands of the earth where population has not yet condensed. With increase of population, these lands gradually come under crops, herding being a transitional stage between virginity and fruition. The food equilibrium of the world turns increasingly in the direction of planting. Even in herding there are epochs ; the present sources of frozen meat upon which the denser populations of the earth now rely for their protein were originally developed not for the meat so much as the wool and leather. Before the advent of refrigerated transport sheep were bred for their wool and cattle for their hides: meat was a by-product to be rendered into tallow before putrefaction had gone too far. To-day the tanneries and wool-sheds are subsidiary interests beside the freezing plants: artificial fibres and rubber will make them increasingly so. Unsupported by its by-products meat will depend for continued production upon a lack of competition by alternative forms of fat and protein. It will not be lacking however; fats are already available from the by-products of cotton—the very crop that will replace wool and leather.

The production of meat, it appears, therefore, will very probably cease unless the food yield from a given area can, by breeding and feeding, be brought to a point when it passes the food value of the 35—40 bushels of wheat obtainable from the acre under careful cultivation. Furthermore, as a source of protein, it can be replaced by wheaten bread, fish and some of the vegetable oil seeds after extraction of their oil. For example, soya bean, besides providing the Western world with soap-making oils, provides the Far Eastern populations, notably of Manchuria and Japan, with various foods very similar in composition and taste to milk and cheese ; and already experiments are being made to devise a means of eliminating the toxic ingredient gossypol, from the protein of the cotton seed. The provision of both oil and protein by land crops will probably render futile the present researches being undertaken to conserve and increase our supplies of fish ; the intensive oceanic cultivation, predicted for the future, which would emulate the oyster breeding and whaling activities of the present day, is never likely to come about. The world will never see its oceanic littoral a chain of fish farms as continuous as its present chain of holiday resorts, with the consequent confinement of the annual seaside holiday to the cliff top, whilst keepers and poachers war beneath on a beach jealously guarded by the fisher-farmers.

For carbohydrates the world depends to-day on a number of crops, the chief of which are wheat, rice and sugar. Statistics of recent years point to an increasing consumption of wheat on the part of previously rice-eating peoples, since the increase is more than can be accounted for by the growth of the naturally wheat-eating population. Hence, if Eastern peoples continue in their adoption of Western culture and habits of life, one may conclude that wheat will gradually become the predominating food crop of the world. The greater vitality of Western peoples compared with Eastern can, with some justification, be attributed to the gluten content of its wheaten staple of diet. Gluten provides the protein requisite for body building, and its absence from a rice diet explains, according to experts, Oriental lack of physical stamina. Providing, as it does, both protein and carbohydrate, fat and vitamin, wheat is, literally, the staff of Western life.

We can increase our supplies of wheat either by increasing the acreage allotted to it, or by increasing the yield from that already under cultivation. Of the two the latter will probably be adopted later, when all suitable virgin land has been taken up, when it will become a necessity. The obvious method of improving wheat bearing land is by fertilizing artificially: by this means Denmark obtains an average yield of 41.8 bushels from an acre. Without it South Australia obtained in 1898 an average of 7; in 1898 the average yield of the United States lands was 12 (the yield established at Rothamsted Experimental Research Station for unmanured land is 12.6); with the aid of fertilizers the United States yield has now climbed to 17. The ingredient of the soil that the wheat plant removes most and which, therefore, must be replaced in greatest quantity is nitrogen; in the provision of nitrogenous manures chemical science has already made an important contribution to life and health. In 1926 the chemical factories of the world converted 1,245,000 tons of atmospheric nitrogen into solid fertilizer, all of which went to the treatment of cultivated land; it is expected that at least half as much again will be dealt with in 1927. With regard to the increase in the yield of cultivated wheat, it has recently been discovered that chlorides have an important influence. Nitrogenous fertilizers of this kind produce a greater number of grains per head in barley than when used in other forms such as the sulphate. Again the number of heads per plant may be increased by supplying the nitrogen in the form of cyananide. Hence a combination fertilizer embodying both these principles should possess extraordinary potentialities in increasing the yield per acre. Because nitrogenous and other manures can only be absorbed by the plant when presented in a limited number of forms chemical research is being directed to the discovery of the most economical molecules to synthesize ; and it now appears that they work most efficiently if due attention is paid to the requirements of the soil bacteria ; these require for maximum activity that the soil should be in a certain physical condition. This condition is achieved in virgin soils by the humus, the top layer of decayed organic matter ; this is progressively destroyed by cultivation and has in the past been replaced successfully. Because of the replacement of the bulk of farm animals by mechanical traction farmyard manure is becoming exceedingly scarce. Malthus remarked in his *Essay on the Principle of Population* that "cattle seem to be necessary to make that species of manure that best suits the land." But Malthus was wrong, the essential is the straw ; it is in rotting this that animal manure is valuable. Obviously, chemistry should be able to provide a rotting agent. Already it has been found : research at Rothamsted has revealed a method whereby any kind of straw may be partially digested to yield an excellent top dressing ; thus yet another crop by-product will replace an animal by-product formerly considered essential.

Before and since Pharaoh's seven lean years growing plant and ripening fruit have been subject to destruction by parasites ranging in size from micro-organisms to grain-loving birds. Direct action against them is always possible, but not so invariably attended by success ; the smaller the depredator the more difficult its extermination ; and the greater chance has prevention rather than cure. Signs are not lacking that in preventive methods chemical science will find increasing employment ; principally in supplying elements to the soil to render the plant immune to parasitic depredation ; it is by attack through the plant itself that most success is obtained. Mangolds, for example, which have been heavily manured with nitrogen and unsupplied with sufficient potash, suffer from leaf spot ;

chemical correction of the deficiency will yield normal mangolds from the same land. Chemical treatment of the seed before sowing is also a successful preventive measure : wheat seed may thus be immunized against smut, a most destructive micro-organism.

Similar treatment that would render the plant poisonous to animal, and therefore human, life, obviously cannot be adopted to preserve crops against the animal depredators that may be expected to multiply at the same rate as the cultivation of crops : these will have to be warred against in much the same manner as at present ; although it is possible that, their size lending an element of sport to their chase missing from microbial hunts, they will provide an alternative to fox-hunting, ruined by lack of horses : the re-introduction of falconry may well be the outcome. Whether football enthusiasts will find summer relaxation in hawking obviously depends upon adapting the principles of greyhound racing to hawk swooping.

Evidently, then, we are already busily preparing for an increasingly intensive cultivation of the earth. Whether we shall eventually cover all the lands of the earth with crops must depend upon our skill in both irrigation and the breeding of plants to thrive in climates differing from their native habitat ; in the latter of these we have already achieved something, as in pushing the Canadian wheat belt further north, but much still remains to be done, particularly in the effect of soil upon making the plant more hardy ; irrigation, too, as Dr. Bernard Keen has shown, presents many purely chemical problems concerning the effects of evaporation and drainage upon the resulting soil. We shall probably find that tropical agriculture will be confined to such carbohydrate and fat-bearing crops as sugar, soya-bean and cotton, and that the more temperate zones will be devoted to the cultivation of carbohydrate-and protein-bearing crops like wheat ; citrus and other valuable vitaminic fruits will be grown in the sub-tropical zones that favour their growth.

The problems, when we have ploughed the whole world, will be, not to make plants grow, but to grow plants bearing the maximum of food. For, although we now modify many of Malthus' conclusions, yet population continues to expand, and, as pointed out, even were the whole world to become a garden, the multiplicity of Adams and Eves would still require to be fed. Whether food supplies would always be so restricted, or tend to restriction to the subsistence level[1] must be regarded as very doubtful. The vast improvements that may be expected through increased efficiency of both the mechanical and chemical processes connected with food may, and probably will, counteract the opposite tendency. Some day it may be necessary to strike a mean between the acreage devoted to food crops and to what may be termed utility crops, although by

[1] In a properly organized world it would presumably be something like the Voigt ration.

then improvements in by-product utilization may well render differentiation impossible.

The conjunction of the findings of dietetics and the possibilities of plant-breeding may well lead to the evolution of some grain never imagined by the wildest of prophetic fantasy. Possibly by the improvement of wheat. It provides carbohydrate, fat, protein and vitamin and it has recently been shown that cross-breeding can vary the relative amounts of these components. By the application of hybridization and Mendelian selection to wheat it is possible by the use of a parent—itself unsuitable for commercial production in one climate—(say England)—to evolve new varieties possessing one or more characteristics of great excellence from the otherwise unsatisfactory parent. An example is the introduction of Red Fyfe wheat into England in 1902 and the breeding from it of many wheats with greatly increased protein content. It will be possible to grow carbohydrate, fat, and protein on the same stalk in the correct proportions for the perfect diet, and when the whole earth is tilled and sown, that will be the ideal of the planter.

The increased utilization of the tropics to make the world a garden of fruit for the varied purposes of man involves the elimination or, perhaps more correctly, the suppression within reasonable bounds, of disease in all its insidious forms. Disease, however, is not confined to the tropics ; it is apparent in some form or another in every part of the world ; but its very universality makes easier the task of its defeat, since knowledge gained under certain conditions may be applied in others less favourable to its acquirement.

The future will inevitably witness the accomplishment of a revolution in our treatment of disease, as complete as the *bouleversement* that followed Pasteur's epoch-making discovery of the microbial nature of contagious afflictions. Because Pasteur showed each disease to be the multiplication in the body of some individual organism, medicine, taking the obvious line of direct attack, has confined its activities to the introduction into the body of preparations, vaccines, anti-bodies, and so on, calculated to poison them directly; medicine in the past has viewed disease from the standpoint of the micro-organism. Now it appears as if medicine will regard it from the point of view of the patient and instead of injecting him with anti-bodies prepared in a foreign medium, will assist him to manufacture his own anti-bodies. It obviously is the scientific method of attacking the problem: to assist the whole organism of its own accord, to set its parts in order, for the body itself to turn its equilibrium from the hydrolyses of death back to the syntheses of life. It is unlikely that we should have discovered it had we not suffered the disappointments and uncertainties by which an insufficient knowledge forced us to learn.

Building on the foundation laid by

Pasteur, we have discovered that the germs of disease, no matter of what kind—diphtheria, malaria, syphilis, or any other, as they thrive, throw off poisons, the toxins, to poison the body and change the vital equilibrium in the direction of the hydrolyses of death. These toxins are extraordinarily virulent: for example, diphtheria bacilli produce a toxin so virulent that 1-25th of a drop of it, when injected into a guinea-pig weighing 500 grams, kills it in a very short space of time; that is to say, a few germs can yield so toxic a poison that the whole internal chemical equilibrium of a mass of living tissue, 250,000 times heavier than itself, can be so upset that death ensues.

Disease having been discovered to be a bio-chemical phenomenon, chemists and pathologists have achieved between them most wonderful results in the invention of compounds to fight it: for logical reasoning and indefatigable experimentation no other activity of human thought has a parallel. Although we shall probably find in the future that their solutions were mere palliatives, yet we shall be indebted to their work for every advance we are likely to make. A path that was tried, first by Jenner on smallpox, was the infection of animals with the disease in order that they should develop in their bodies the anti-toxins to fight it ; preparations of these could then be taken and injected into the blood of similarly infected patients, providing them with ready-made anti-toxins.

Both these methods have yielded significant results which, pooled, show that (a) under healthy conditions the body generates the necessary anti-toxin when attacked by disease ; and that (b) the anti-toxins are peculiar to the disease, as are all catalysts, whilst (c) that illness is the stage when the body's anti-toxins have proved insufficient to resist attack completely. These are the facts that chemo-therapy will work upon in the future. The task is to discover the necessary excitant that shall stimulate the diseased body to produce its anti-toxins when disabled by disease. Already some progress has been made : the body's manufacture of anti-toxins, in all probability, is localized in a depôt from which the disease-destroying chemical is distributed as, and when, required by the blood, through the body to the site of infection. Two bio-chemists have been able, during the last few years, to form such depôts in rabbits, to protect them against syphilis, by using insoluble bismuth compounds ; such depôts heal up when no longer required. Evidently if only the suitable excitant is provided and a depôt formed, the body will cure itself.

This depôt-forming habit of the body only just correlated with disease, has, however, been recognized for some time as its method of maintaining its normal healthy functions. For these it requires the services of, and it manufactures, a number of exceedingly complex substances known as the hormones, which it makes and stores in the ductless, or to give them their bio-chemical name, the endocrine glands. These hormones are among the body's own catalysts for the syntheses of life. Although each hormone is apparently specific for certain tasks yet they are all interdependent: failure of one lowers the metabolism of the whole body and may even lead to its death, as in the case of the failure of the hormone, insulin, of the Islets of Langerhans in the pancreas, lack of which produces diabetes, a sugar poisoning of the blood, leading speedily to the poisoning and death of the whole body. So far our knowledge of the hormones is confined to only a few but already it has enabled us to banish certain types of abnormality, such as cretinism, a disease, chiefly of young children, caused by congenital absence of the thyroid gland, or by insufficiency of its secretion, resulting in stunted body and mentality, and a thick, flabby skin. Research has progressed so far that we now synthesize in the chemical factory, in quantities that, if called for, could be expanded practically illimitably, three extremely important hormones: thryoxine, the hormone of the thyroid gland; adrenaline, of the suprerenal gland, and histamine, a hormone found in the wall of the lung. Others, knowing less surely, we can nevertheless extract from similar glands of animals: a side-line of the slaughter-houses of the world is, to-day, the extraction of valuable drugs. It is, one may hope, only a matter of time before we can synthesize them all and banish every kind of abnormality, both mental and physical, from the human body. Then the path will be clear to the perfecting of the race.

Before we are in a position to achieve this, we must first learn much more about the vitamins, those elusive super-catalysts without which all the vital syntheses appear impossible. So far as we have investigated, we have found evidence of five and, because of our poverty of knowledge concerning them, we can indicate no more of their chemical composition than is embodied in the names we have given them: vitamin A, vitamin B, vitamin C, vitamin D, and vitamin E; a vitamin X has often been found, but further experiment has always proved it identical with one of the five. We know nothing more about them than the facts, of their existence, that lack of them produces disease, and that the human body appears unable to synthesize them, or to store them for very long. Lack of A prevents growth, of B causes beri-beri and polyneuritis, of C produces scurvy, of D produces rickets, and of E impossibility of reproduction both in male and female.

So far they have entirely eluded chemical analysis, and therefore, synthesis. Vitamin D is the one hopeful exception. Its preparation can scarcely be called chemical synthesis, as it has been made, or, rather, a resultant has been obtained, giving similar effects, by exposing a complex substance, called ergosterol to irradiation by ultra-violet light. This vitamin appears to be stored in the liver, for the short time it is stored by the body; possibly the other vitamins may be stored in the glands of the body.

In the vitamins and hormones we appear to reach up as far as we have yet been able, towards the controlling factors of life. Possibly, when we have examined them fully and prepared them, we shall be within measurable distance of a synthesis of life : perhaps not of any form of life we already know but a rudimentary cell that conceivably might, through successive ages, evolve through the same stages we ourselves have passed. Any hope there may be of synthesizing an articulated, complete, free-thinking human being must be dismissed as beyond all possible hypothesis. It is definitely impossible ; the labour alone of putting the parts together, even were they made, and that *might* be possible, would be too infinitely intricate a task.

It may safely be predicted that Nature's method of maintaining the succession of the human species will remain the most economical and only practicable one, even though one may say that it will certainly be profoundly controlled and consciously guided. The possibility, however, of the parthenogenetic propagation of the human species by the surgical removal of an egg from the ovary and its subsequent fertilization and culture in synthetic media as advanced by some, particularly Mr. J. B. S. Haldane, in his *Daedalus*, takes far more for granted than is likely to be accomplished. The first essential must obviously be a thorough knowledge of all the vitamins and their influence both on mother and fœtus; and secondly, we should require a knowledge of the effects of the hormones of the mother upon the fœtus. These are not impossibilities, and were they the only factors involved the process might be considered vaguely feasible; it is rather with the consequences to the race of the results of the process that make one sceptical of its possible adoption. Mr. Haldane advances in favour of such propagation the saving of trouble, energy, and pain to individuals of both sexes, that it would bring with it and the consequent freedom it would confer on them for mental activity and inquiry. One must beg leave to doubt the possibilities of mental activity in view of our knowledge of the results of a total deprivation of gonadal activity, particularly in the male, who, presumably, would be most affected.

The chemical control of fertility will probably be the first step, since recent research indicates its government by one, or perhaps more, of the vitamins. Vitamin E has been proved essential for the growth of the mammalian embryo. Lack of Vitamin E appears to react differently in the two sexes ; in the male, if continued over a sufficient length of time, it results in the total destruction of the male gonad, or testis, which subsequent administration of it has no power to repair. In the female the consequences are not so serious, damage is reserved for the fœtus and not the gonad, since vitamin E appears to be essential for a certain hormone made by the ovary in preparation for pregnancy, and, therefore, a necessity for the growth of the fertilized germ. There should be little fear of an abuse of the power this places in our hands in view of the widespread knowledge of the importance to the whole body and to the mind of all the secretions of the body and, particularly those of the genital organs.

Further, such self-deprivation of sexual potency, one may safely conclude, would be the last act of the decadent. The knowledge does, however, provide an alternative to the methods of surgical sterilization of social undesirables that is practised by certain governments. Rather than abuse, it is likely to lead to a fuller appreciation of the needs of pregnancy and, hence, a better care for prenatal conditions.

The chemical control of sex cannot be regarded as remote even though there appears to be no indications of its introduction in the immediate future. When it is made possible, civilization may well have to confront a danger of real gravity. At present there is, and so far as history can show, there always has been, a preponderance of women; not only are more female children born, but they weather the ills of childhood better, and they live longer. The female is statistically the stronger sex, and may attempt some day to change the balance to her own social and economic advantage. Modern anthropology points to the necessity of a numerical preponderance of females for the perpetuation of a civilsation. If this natural order is reversed, either consciously or unconsciously, and not allowed to right itself in the course of time, we may well fear a decline of our race and the end of Western civilisation.

The development in the child of a full complement of hormones will be an increasingly attended care of the future. The early combat of abnormalities may, in extreme cases, take the form of chemico-surgical development in the child of its ductless glands, or it may even be found possible to stimulate dormant glands (and, in passing, it may be remarked, that infantile deficiency of a gland may turn out really to be the result of wasting through lack of stimulation of a gland dormant at birth) by administration of a specific catalytic stimulant. To prevent cretinism, for example, the giving to the young child of some form of thyroxine may be found capable of stimulating the thyroid gland to activity and initiating its normal healthy development. The pre-natal care, already mentioned as a probable future development, will be followed up during the later years of child growth. The exact machinery for achieving this is perhaps outside the chemical sphere, but it does seem that some modification of our present system of Infant Welfare Centres is more likely to be the solution provided by the future for child-rearing problems than any of the predicted systems of communal incubation which, after all, seldom go sufficiently far. Provided the child is normally healthy at birth, there seems no reason why it should not survive the first few weeks of life. It is through the remaining weeks up to the age of five years that, from statistics, care is most needed for future growth, hence, since most prospective schemes appear to provide only for a brief incubation after birth, they would not be likely to attain their objective. Thus through completer knowledge of the chemical control of the body it will become possible to evolve a physical and mental type finer than any previous generation has ever dreamed. Thus it appears that we may anticipate a great evolution not only in material things and methods but also in man himself.

Geber, the delineator of exact method, Hermes, the beckoner towards more daring synthesis, are at last beginning to enter upon their kingdom.

SIBYLLA

OR

THE REVIVAL OF PROPHECY

BY

C. A. MACE, M.A.

Lecturer in Logic and Psychology at the University of St Andrews

LONDON
KEGAN PAUL, TRENCH, TRUBNER & CO., LTD.
NEW YORK: E. P. DUTTON & CO.

Printed in Great Britain by
STEPHEN AUSTIN & SONS, LTD., HERTFORD

FOREWORD

Can Prophecy be a science? Science, at any rate, appears to aim at prophecy. We are often told that the test of an hypothesis lies in the events that it predicts; but it is a test that is much too rarely applied. We are surprised when Biologists apply it in a systematic way to the theory of evolution. Historians also tell us that the study of the past will help us to foresee the future, but in their practice they hardly succeed in catching up to the present.

The organum of human knowledge, in fact, presents a curious spectacle—a vast system of foundations but no sign of the edifice these foundations are to bear. At least such was the case until recently. In the little prophetic volumes recently published we can see, perhaps, the preliminary sketch of its imposing elevation.

And so another question occurs. What is the future of prophecy? Psychical research may even yet surprise us. Latent powers of divination may lurk in the human mind, but it is premature no doubt to look for enlightenment here. Less sensational but more significant is the emergence of the prophetic function in the work of science itself, and in the pedestrian progress of our general intellectual life in the twentieth century.

It was this prophetic function of scientific thought which the author set out in this little book to investigate, but like other writers in the series he was tempted to "try out" some tentative principles which seemed to promise well. Perhaps the future will reveal some inaccuracies in his forecast, but the digression has been worth while. One thing, at least, is certain. Large scale organization extends to the work of science. In our past experiments we have been too much concerned with events of short duration, with the kind of results we could obtain for rapid publication, and with cycles that fall within the span of an individual life. Even now we have insufficient data to decide whether cold winters come in periods of seventy or eighty years, a ridiculously easy problem which a little organization would have solved. But every year will add to the fund of carefully recorded data. Men are learning to co-operate through time. Experiments initiated by the father are continued by the son, and succeeding generations will record the results.

Organization provides a workable alternative to the invention of Methuselahs; though it would simplify things considerably if Nature could see her way to adopt the suggestion of Mr. Shaw. More than anywhere else perhaps this kind of organization is needed for sociological and biological experiment. By that way we shall obtain the knowledge upon which scientific prophecy may be based. In the meantime, let us speculate. Speculation is not a deadly sin, but an aid to true perspective, and it is suggestive of good hypotheses. Nor need our speculations, even now, be wholly in the air. We have the broad tendencies of the past on which to base our forecast. The danger lies in a premature attempt to elucidate the details. The older prophets tried to foresee particular events, the newer are content with the general tendencies. Knowledge grows not like a crystal by minute accretions, but like a work of art, from outline down to detail guided by inspiration—which, in science, is speculation.

C. A. M.

SIBYLLA

OR

The Revival of Prophecy

I

Experience enables us to defend in age the prejudices of our youth, and belatedly to realize ambitions forsaken in earlier years. This would seem to be true not only of individuals but of institutions also.

Modern science, adolescent in the nineteenth century, shows signs of middle age ; and with it that mellowed and urbane tolerance which fosters idealism and optimistic projects, projects which the realism and studied pessimism of youth declare to be beyond all powers of attainment.

The researches of Rutherford and his fellow-workers have disposed the serious physicist to dally once more with the Philosopher's Stone. Again we hope to transmute the baser metals into gold. Indeed, the Philosopher's Stone is found. Japanese physicists claim that a magnification 2,500 times of the produce of their experiments enables us actually to photograph the world's augmented gold-supply. Furthermore, the progress of knowledge concerning atomic structure and radiant energy promises to endow man with the powers and range of influence for which the magicians of old had such a notable though undeserved reputation.

Since Freud, the interpretation of dreams has become a serious preoccupation of the grave and a further relaxation for the gay. I need not, however, multiply examples. Let us turn forthwith to the latest phase in this development — the revival of prophecy.

Prophecy, of course, has never wholly died. Though the soothsayer fell into disrepute and was banished from the temple precincts to earn a precarious livelihood on Margate Sands, his mantle has fallen on Mr Wells. Mr Wells, however, is merely a prophet of the transition, the journalese precursor of a new and hardier race whose influence in the world I suspect will prove much greater. Like the modern prophets generally, he has substituted for revelation and intuition a thorough-going reliance upon the mundane intellect, fed upon scientific generalizations and historical particularities. Mr Bernard Shaw is another of the transition prophets, having marked affinities to the new Biological School; but, like the rest, is transitional in lacking the regalia of office and authority with which the newer prophets are endowed.

The first of the prophets of the twentieth-century school was Mr J. B. S. Haldane. In *Daedalus* we find a frank abandonment of the pose of scientific reserve about the future. Scientists generally for long had said that they were concerned only with verifiable facts. Newton, in particular, had said quite publicly that he had no use for hypotheses; and his successors for the sake of their reputations had to keep up the Newtonian manner. To distinguish themselves from journalists and philosophers they avoided speculation like a plague. Now Haldane found himself, I imagine, somewhat unduly constrained by this taboo. Pursuing the most rigid methods of inquiry, he had come to conclusions about some probable developments in the future, conclusions much less speculative than many of the things it was his duty as a biologist to teach about the past. So he prophesied.

The relief in the scientific world was heartfelt and instantaneous. Scientists and others, of various degrees of eminence, literally scrambled into Mr Haldane's bus. In the year or so that followed some twenty prophetic volumes had been produced, and the publishers tell us of many others in the press.

However, in many of these documents we fail to detect the authentic guiding hand, and these uncanonical works we must leave to the Higher Critics of the future. The genuine prophet is to be known by his strictly impersonal manner. He tells you neither what he wants nor what he fears. So far as we can judge, he has no wants or fears. He merely tells you what is going to be. Judged by this test we must, I think, reject *Thrasymachus*, as purely propagandist. Mr Joad wants to popularize immorality. He tries to persuade us to join the ranks of the libertines lest we be swamped in the coming Neo-puritan revival. By the same token we must, reluctantly, reject Mrs Bertrand Russell. In *Hypatia* the quasi-prophetic rôle is assumed, we fear, merely as a vantage point from which to wage sex-war.

Coming to the genuine prophets, I would divide them into two kinds; the mechanists and the vitalists, not however, using these terms in their traditional philosophical senses. The mechanists are those who view the future in terms of the development of machinery. Man, having a nature very much what it is to-day, is seen in an environment of mechanical perfection, a world of wireless telegraphy, television, fuelless traffic, moving pavements, rubber roads, windows of flexible glass, rustless metals, dustless and smokeless cities, and private houses fit for a plumber's paradise.

The vitalists, of whom I take Mr Haldane to be the leading spokesman, have a message in the face of which the wonders of wireless are relatively tame. The time has come, they say, when our science and our inventive genius are to be applied to life itself. In a sense, of course, the journalists and men of letters got in first. Kapek with his Robots made the idea dramatic ; but the mechanical tradition was too strong. The Robot was little more than clockwork. Wells's " Men like Gods " are perhaps a slight advance, but the gods are all too human, and a bit too much like Mr Wells. Shaw, I think, got nearest. In *Back to Methuselah* the problem of biological invention was approached in a more biological manner. But all these are prophets of the transition. They are at heart Utopians, and neglect the evidences of the present upon which the forecast must be based.

II

It would tempt us too far afield to inquire into the logical foundations of the new science of "Prophetics", as no doubt this branch of learning will be called. A young science must not at first aim at too much systematic order, nor be too introspective. Let us, rather, glance into the future and try to fill in the larger of the empty spaces in the vision so far presented to our eyes.

We can readily grant at the outset

all that the mechanists claim. In fact, there is nothing, I think, unlikely in the suggestion that, so far as mechanical invention is concerned, we are approaching the time when all that is physically possible will be realized. The future customer at Selfridges' who inquires for a teleofactor in order to smell the perfumes of Arabia will be told that the impossibility of this has now been finally proved. For the rest of his desires, however, he will be quite amply supplied. He will have perfected wireless concerts and tele-visual cinemas; he will be conveyed by the morning aerial Pullman from his villa-palace in Devonshire to his office in what was once the Strand. To his Club or restaurant he will proceed by a moving pavement, or in a bath-chair propelled by radiant power. His week-ends may be spent in Samarkand or in tobogganing on Mount Everest. His wife, too, will benefit by this universal progress. Her day's work will be done by 9 a.m. The turning of a tap or two will nourish the ectogenetic child foretold by Mr Haldane, and the pressing of a button will put in action the automatic cook. By wireless communication with Paris she will choose a garment, to be delivered by an aerial messenger boy upon his cycloplane. After luncheon she will, again by wireless, put herself in visual and auditory contact with other members of the society to which she belongs. It will probably be a society for the revival of twentieth-century customs.

But, granting all this, the obvious question occurs. What will be its effect upon human nature? Is it conceivable that the perfecting of machinery could leave it unaffected, and will not inevitably this inventive genius be turned upon man himself? Our future man of business who, by improved modes of locomotion, spends his week-ends in Tenesee, Kikuyu, or Zanzibar (according to his religious predilections), will he enjoy only the advantages of mechanical invention? Surely a perfected Pelmanism will enable him to learn three native languages on the Friday's journey out, and an improved process of repression assist him to forget them on the Sunday evening's journey home.

This is not entirely idle speculation.

Applications of science to the control of the human mind have already begun. They started, perhaps, in the crude and blundering experiments in Scientific Management. They are groping towards more solid foundations in Industrial Psychology—the philosophical significance of which has, up to the present, escaped the attention it deserves.

Consider briefly the course of this movement up to the present date. Sometime in the eighties Taylor introduced his principles of Scientific Management. Some ingenious spirit, indirectly inspired no doubt by him, produced the following invention for organizing the activities of man. A certain American establishment paid its employees on the basis of 'task work', i.e. upon piece rates with a time-limit for the performance of the unit-task. The worker was paid a 25% bonus on the ordinary wage for the performance of his task, but the task and time-setter was paid a bonus based upon the number of men who failed to earn their bonuses. The apparent disadvantage to the worker, however, was counterbalanced by providing him with a foreman who also was paid upon the basis of the number of men under him who 'made their tasks'. As Muscio [1] comments, "The situation then was this: The workman was given a bonus as an incentive to expend intense efforts to accomplishing a task in a set time; the foreman

[1] Muscio, *Lectures on Industrial Psychology*, p. 39 (Routledge).

was given 'blood money' to drive the man if he became slack, and the task and time-setter was also paid 'blood money' to set the times so short that 'the making of the tasks' involved an expenditure of more than the greatest reasonable amount of energy." The Trade Unions got to know of this, with the result that Scientific Management of that sort is now as dead as mutton. Public interest subsides. But in principle it was right—right in the sense that its inventor had a glimpse of something which is ultimately going to prove effective. He made the mistake which probably every inventor makes. He constructed a machine with the power to blow itself up; but this only shows that power is there, and power to be controlled. The mistake, of course, was a natural one to make for people preoccupied with the parallelogram of forces. Another case supplies the necessary comment and indicates a more hopeful line of experiment.

" There was in operation for some years at the leper colony off the Philippine Islands a system of weekly gratuities to each man, woman, and child confined to the island colony. From the women and children no accounting for the subsidy was required. But from the men a certain amount of manual labour about the island was exacted upon penalty of having the pocket-money withheld. From the administrative point of view this had seemed an easy solution for the difficult problem of getting adequate labour in an isolated place inhabited largely by the victims of a dread disease. But the men patients took vigorous exception to this form of compulsory labour, and finally made complaint about it to the Philippine Government. An investigation into the unrest at the leper colony was instituted, and the Secretary of the Interior visited the island and heard all the complaints in person. As a result of his study the system of gratuities for the men was wiped out. And the necessary work on the island was paid for at an agreed rate which, it appeared later, was less than the previous gratuity. Nevertheless, the men found the new system preferable; there was no more complaint, the necessary work was done; the men who were inclined to work received their stipends and the others did not. But from that day to the present trouble on this score has been unheard of." [1]

More systematic were the investigations in this country made under the Ministry of Munitions. In the Interim report of 1917 of the Health of Munition Workers Committee we find the following reference to incentives in industry.

"It is essential that the wage system should be equitable and easily understood by the workers. The evidence collected leaves no doubt that a wage system the operation of which cannot easily be understood by

[1] Ordway Tead, *Instincts in Industry*, p. 1.

the wage earners . . . fails to serve as an incentive."

In fact, the evidence in detail rather suggests that it is more important for the system to be intelligible than it is for it to be just. As in the case of the Philippine lepers a reduction of wages may in certain cases be the shortest remedy for industrial discontent, just as the reduction of working hours has often been found to involve an increase in production.

More important, however, than juggling with the systems of remuneration is concentration upon the subtler psychological factors in the situation. Politicians, artists, bishops, and many members of the working-classes are played upon by forces which act more powerfully on their energies than any system of wages that could be applied to them; the ambitions and interests of their wives, the health of their children, the criticism or adulation of their associates and a host of other factors of curious psychological interest. Is it beyond the wit of man to apply these forces in a scientific way?

One very instructive attempt was that of another American establishment, which instituted the office of 'plant mother'. "It is her duty" (as Tead puts it), "using this motherly disposition and attitude as an entering wedge, to go among the men and help to straighten out their troubles with the management. And one of the most successful weapons of appeal with her is said to be that she puts employées' problems in family terms. For example, a man will want to quit because of a slight ruction with a fellow-worker or a superior, whereupon she will remind him that his little Johnny should not be forced by father's unemployment to leave school to go to work, or that another baby is coming in a couple of months and that he mustn't cause anxiety to 'the wife'. . . . This particular 'plant mother', I am told, has been instrumental in reducing the labour turnover to an astonishing extent." [1]

My readers will be able to think of many other possible applications of this principle. Cannot our domestic parrot be turned to economic advantage? Its present function would

[1] Tead, *op. cit.*, p. 28.

seem to be only that of preserving the unity of private family life by giving its jaded members something to laugh and talk about when the bonds of common interest and filial piety are beginning to wear thin. That which preserves the family from disruption and armies from revolt will surely soon be dedicated to the cause of peace in industry.

To these things we are gradually and insensibly being led, so gradually and insidiously as to evade the inhibitory sense of the ludicrous. Already it is known that light, ventilation, and the general comfort of the worker are relevant to production. By slight and inexpensive changes in these respects output may be augmented say 3·5 or even 10 per cent, and the worker suffers less from nervous irritability and is less punctual in paying his subscription to his union. At any moment it may be found that hanging pictures on the workshop walls has a similar effect. Possibly Landseer's would raise efficiency and contentment only ·1 per cent, whilst Sargent's would really justify their auction prices. We should then discover a more satisfactory policy with regard to our National Galleries. We should go to Port Sunlight to see good pictures and the Royal Academy would become the testing hall for experiments in industrial art.

All this is coming about no longer by mere chance or blind experiment, but by the operation of a principle which I take as a foundation stone for prophecy. Man has slowly and painfully acquired facility in adopting the scientific attitude of mind. Gradually he has been able to achieve this point of view in relation to the material universe. But he finds a peculiar difficulty in being scientific in the face of the animate; and the difficulty would appear to be almost insuperable in the face of mind. It is because we are living at a time when the feat is being accomplished that I feel impelled to prophesy an important revolution, a revolution beside which the Bolshevic incident, for instance, is dwarfed to almost comical insignificance.

The revolutionists are level-headed but far-seeing men of business, acting in alliance with the scientists. Here is their programme as voiced by Professor Cattell in 1903.

"It is our business to make both a science and an art of human nature. As in the physical world we select first the material suited to our purpose, and turn the iron into steel, and temper the steel for a knife, so in the world of human action we must learn to select the right man, to educate him and fit him for his exact task. This indeed we try to do in all our institutions, religions, commerce, system of education, and government. But we work by rule of thumb—blind, deaf, and wasteful. The nineteenth century witnessed an extraordinary increase in our knowledge of the material world, and in our power to make it subservient to our ends; the twentieth century will probably witness a corresponding increase in our knowledge of human nature and in our power to use it for our welfare." [1]

Let us note the implications of this programme. At present we have made but a few experiments and a few discoveries in industrial psychology. The future, however, is to see the principle applied to education, criminology, government, war, perhaps religion and the intimate personal life of man.

Can we foresee the consequences? To some extent, I think, we can. As Dr Fournier d'Albe puts it in the

[1] Homo Scientificus Americanus: *Science*, April 10, 1903.

first of his prophetic volumes,[1] we have but to extrapolate the curves of present developments into the future. We must, however, also take into account so far as possible some of the other factors at work. The general course of evolution is in outline known to be one of differentiation, specialization, integration, growing consciousness of ends, and substitution of systematic experiment in place of blind trial and error. We shall find other clues which, carefully studied, enable us to speak about the future with at least as much assurance as biologists and anthropologists enjoy in speaking about the remoter past. We shall, moreover, have the added benefit of finding confirmation or correction in the years to come.

[1] *Quo Vadimus*.

III

Facts brought to light in the sphere of industry have an application in politics, so to politics and the arts of government let us turn. Let politicians consider, for example, the discovery that the incidence of shell-shock in war, and of nervous disorders in peace, is highest in the dangerous occupations which are also the occupations most liable to industrial unrest. Such facts prevent us from viewing the discontented worker either as a martyred idealist, or as a mere criminal. The case is paralleled by the suggestion that desertion and malingering in exhausted armies are symptoms or 'defence reactions' in incipient shell-shock. They force upon us a more dispassionate and a more therapeutic point of view. The criminal in general is coming to be regarded in this light. Turn now to the anarchist, and the psycho-analyst explains to you that his activities are minor symptoms of his 'father complex', which imprisonment, for example, will merely intensify.

From knowledge of the conditions of these things proceeds control. Possibly punitive measures may be genuinely counteracting causes. Perhaps, however, empirical science may declare that the best treatment for an anarchist is to make him a Lord Mayor, and the best remedy for a militant labour party is a term of office judiciously supervised by our permanent officials. The future student of social affairs will be preoccupied with the scientific investigation of the sources of power and its social reservoirs. The 'party in power', as it is charitably called, has not much of it. Public opinion seems to have much more. In fact, the chief function of a government, now-a-days, seems to be that of carrying out the programme of the opposition.[1] The opposition is relatively free from criticism; the Government moves in continual dread of it, being threatened by attrition in its bye-elections and by a general landslide at any appeal to the country.

Under such conditions the essential tasks of government are very much

[1] We have recently witnessed a paralysed Labour Government incapable of carrying out its projected remedies for industrial ills, and a Conservative Government maintaining a system of Free Trade.

the same of whatever political colour the party in power may be. First and foremost it has to deal with emergencies, whether spontaneously arising or organized by the opposition, in such a way as to preserve its popularity with the public. It has also to carry out more or less efficiently the routine work of government, and it has to prepare for the next election by enshrining its most permanent tactical advantages within the framework of the Constitution. Any time and energy that may be left can be devoted to carrying out its 'programme' and in dealing with what it conceives to be the fundamental disorders of society.

A 'party programme', like a dream, has a manifest and a latent content. The manifest content is drawn from the incidents of the day and popular demands. The latent is constituted by those deeper policies for the furtherance of which this clap-trap may be closed. The distinction between the manifest and the latent content is not due, as is frequently asserted, to mere political dishonesty but to the necessities of the case; the primitive group mind's need of a 'Cause' and of an outward and visible sign of an inward and spiritual economic doctrine. With the increased complexity and scientific character which the latent content is, in future, likely to assume political sacramentalism will become more and more important.

How will this affect the issue of our present discontents? Viewing the situation frankly, we find in society two powerful and opposed forces. Let us adopt the convenient labels and call them Capital and Labour. It would take us too far afield to discuss in detail the prospects of success in each of the various remedies suggested for their enmity. In general, it seems to me that all the proposals depend for their success upon large-scale persuasion, and ignore the fundamental causes of unrest and conflict. Even labour, it seems to me, is mistaken about the causes of its troubles, and labour might be expected to know them best. But even to ask this would be as unreasonable as to ask a patient to diagnose and prescribe for his own disease. The solution, I venture to think, will come along the lines upon which the only solution of any problem has ever before been found—along the lines of exact and scientific thought and piecemeal investigation. The recognition of this combined with attention to the apparent power of public opinion in modern politics has led many to suppose that the ultimate remedy lies in the dissemination of education, of exact and scientific thought throughout our great democracies. But, personally, I doubt the possibility of disseminating anything, least of all scientific habits or attitudes of mind. It would seem to violate the rule of differentiation. Nor can I bring myself to believe that there ever was or is ever likely to be a democracy in any important sense. It is sufficient if this scientific spirit be exploited by those who hold the reins of power, and it is because it seems that they are beginning to see the possibilities of this that I think the really significant revolution is to come. What is called 'Public Opinion' is now-a-days a carefully manufactured article produced by firms entirely outside the scope of the Factory Acts. Of course, even a despotism is a 'democracy' in the unimportant sense that the despot's rule depends upon a certain kind of consent. What has happened lately is that a new kind of leader has been evolved. The old leader ruled by certain kind of obvious strength. The new ruler practices an art. He is, moreover, a student of the science of persuasion. He appeals through various media directly to the masses, and does not so much inspire confidence as suggest that he is only a more articulate spokesman of their will. The contrast is rather like that between coercion and hypnotism, two obvious ways of making people do as you wish; and the transition from one to the other does not seem to me to be adequately described as a passage to democracy.

The future, it seems, will see these powers scientifically employed. Much has been learnt by crude empirical methods. The newspapers, the advertising agencies, and the publicity departments have a store of knowledge upon which politicians are wisely beginning to draw. Soon they may attend to Mr Bertrand Russell, who in his prophetic volume, foresees the development of glandular psychology.

We know that by suitable injections a person's emotional mood can be changed, and Russell suggests that a simple operation like compulsory vaccination will convert the reddest Bolshevic to the temperament of the mildest curate. The glands of docility will be under government control.

The suggestion appears to me to be over-simple, and to belong to the mechanistic type of prophecy. Its supreme simplicity is in this connection its supreme defect. The only defence against democracy is a complicated mode of government. If methods of government are fool-proof then any one can practise it; and a sudden *coup d' état*, or an invasion of half-a-dozen uninoculated savages, would be an irrevocable disaster.

It is, I think, mainly through the development of scientifically conducted propaganda that the world may one day see the abandonment of force. No wholesale change of heart will be involved. It will simply be that a more efficient weapon will have been discovered, a weapon that is preventive as well as corrective in its use.

The military mind is, it is generally agreed, conservative to the verge of sheer stupidity, but even the military mind is becoming infected with the new ideas. Under the influence of Captain Liddel Hart, the military representative of the prophetic school with which we are concerned, it will surely come to recognize "the moral objective of war." [1] Even this author

[1] See *Paris, or the Future of War*.

is conservative, in regarding tanks and aeroplanes as being the principle weapons of the future. With regard to the immediate future he may be right. But the implication of his general thesis is that all these weapons will ultimately give place to subtler ones of psychological design. Hardly less effective than the 'death ray' in bending the enemy's will is the rumour that one has been invented. And rumours can be invented daily in His Majesty's Office of Propaganda. And why stop at the control of fear? Why not apply psychology throughout the whole of the enemy's mental life? Perhaps, in the last of all wars to end war, victory will fall to the side which first makes its enemy laugh. The power of a sense of humour, at any rate, is a terrible weapon in the negotiation of peace. Our late enemies would scarcely have fared so badly had they been less sparingly endowed with it in war.

Humour has its applications in peace as in war. Under universal Scientific Management, the Government will vie with the underground railways for the control of public laughter. The Catholic Church, in enlisting the aid of Chesterton and Ronald Knox against its irrefutable but dreary adversaries, has for once in modern history placed itself in the vanguard of a winning cause.

The weapon has curious possibilities, but is one difficult to employ. Quick wits sensitive to the trend of fashion are the essential qualifications for a

Minister of Public Humour. For the joke is the most ephemeral of all works of art. It takes years of patient research, now, to see the jokes of classical antiquity, and pages of solemn argument to defend the taste of Shakespeare in his humour. Differences of humour separate age from age and class from class. It is the weakest link which binds the intellectuals to the cause of labour. One might expect this weapon to favour established institutions, for only the fat and prosperous can laugh with merriment, but by the same token the jokes of Marie Antoinette were unintelligible to the hungry. Therefore the lean and plodding journalists will come into their own and take their place among the expert advisors to the future Ministry of Talents.

IV

Education will perhaps be one of the most interesting fields of future scientific management. At present we know a few bits of a few sciences, and a handful of historical facts. We fill the picture out with theories and blunderingly try to make other people memorize these things. We call the process Education. We have invented a few devices for discovering what they forget, and call them Examinations. Recently we have invented a few tricks of exposing what they cannot do, and we call them Mental Tests. We wrap it all up in a tangle of loose philosophy, and we call it The Science of Education. We profess, it is true, some loftier pursuits. We are trying, we say, to teach our pupils to think. We try to mould their characters, and we try to produce good citizens.

But what does it all amount to in the end? The most successful way yet discovered of teaching people to think is for the teacher himself to think aloud, and hope that his pupils will catch the knack. How do we set about moulding the character of a child? By moral precept, good example, healthy sport, uplifting atmosphere, and an unprincipled application of punishment and reward. Our children are bored by precepts. They yawn in our presence and make fun of our precepts in our absence; they break out at the first opportunity and "sow their wild oats", and then rediscover our old morality for themselves. The more adventurous invent a new one of their own, probably a finer thing than that we tried to teach. Our good examples produce two alternative effects. If the child happens to like us, he will imitate our actions, and morality becomes an empty mimicry. Perhaps, on the other hand, we have an unfortunate mannerism or some innocent resemblance to a forgotten terror, and the child dislikes us. The good example is taken as a pattern of the thing to be avoided. By counter-suggestion he takes the opposite course so far as in him lies. The sons of clergymen might provide us with material for a theory of moral education.

Our healthy sports—but why go on ?

It is not my intention to ridicule the best things that the wisdom of the ages has produced. My only contention is that our methods of education are hopelessly tentative, uncertain in their effects, justifiable only on the ground that we cannot yet see clearly anything else to do. But things will change. Already there *is* a science of education, because people with scientific minds are thinking about the problems it presents. They are experimenting, and they are at least disclosing our ignorance. Experiments similar to those of Scientific Management are producing similar results. We are beginning to know important things about the mind, its natural development, and the method by which it works. Already we know a good deal about the conditions of remembering, enough at least to condemn, if we dared to think out the implications of our knowledge, many venerable educational practices. Some of the most important of our discoveries fall readily into line with the results of Industrial Psychology. There seems, for example, to be a natural rate of working appropriate in learning, appropriate rhythms of impression, rest and recall, by the operation of which knowledge becomes firmly embedded in the mind. When these rhythms are ignored, the evil effects of cramming are produced. Our educational methods in the school and in the university break and crush all these rhythms, and students pass out into the world with mutilated minds.

The rhythms of life, however, present in every field possibilities of extensive modification. That many needs demand a rhythmic satisfaction is well known, but there is scope for detailed investigation and experiment here. None of these natural periods seem to be absolutely fixed beyond all power of alteration. The rhythms of digestion probably vary with the fashions of the age and class, according as high teas or late dinners are in fashion. They might under certain conditions have depended upon the tides; and changes in our habits of taking nourishment would probably produce most interesting variations.

Why should we sleep throughout the night, and work throughout the day? Our habits in this respect were formed in the days before fire was stolen from the gods—and they are wasteful habits, too. Artificial light one day may easily come to be as cheap as water, and man will change his ways of taking rest. Some experimenters urge that the most valuable part of sleep is in the first few minutes. The remainder of the night we spend in dreams and in gradually waking up. Let us apply the principle of rest-pauses in a thorough-going way. Distributed periods of sleep might prove a great economy. People will be taught to recline say once in every two hours, and sleep for twenty minutes. Thus will be inaugurated the twenty-four hour day, and a race of energetic Napoleons will emerge.

But to return to general education.

Why should the period of learning be concentrated wholly in the early years of life? Distributed periods of study will undoubtedly prove more effective, say two hours every day, or two days every week, or, if you like, two continuous months in the year throughout the whole of life. Thus might come the solution to a variety of problems. We have, first of all, the problems of education itself. At present we arrange the life of a child so that a period of intensive cramming is followed by the abrupt cessation of intellectual life at the time when the intellect is just beginning to mature. There is the problem of industrial fatigue, due more often to monotony than to arduous labour. Recent researches afford scientific support to the adage that a change of work is as good as a holiday. Educational pursuits will provide the necessary change. Already many working men spend their evenings at classes and their only holiday at a summer school.

Lastly there is the ever-growing problem of the prolongation of infancy. Under simpler conditions a child was adequately equipped for practical purposes by the age of ten or even younger. Now-a-days fourteen years is considered necessary, whilst the professional man is almost thirty before he is fit to earn a living—and still there is more to learn.

Given successful scientific research in this direction, back we go to the methods of the early days of the Industrial Revolution with children employed in every factory. Under rational supervision it may prove a great advance. The moral education which practical life alone can give will commence at the proper time. Children will be protected from the dangers of neglectful or over-solicitous parents. They will acquire at an early age the much-required sense of responsibility and independence, and their parallel work at school will acquire some semblance of significance and interest.

In addition to new rhythms and periods of study, new methods will be employed. Probably the curriculum itself will be the first thing to be revised. In spite of a good deal of relevant knowledge, no one has yet consistently thought out an answer to the question: What should a child be taught? Freed from the assumptions of faculty-psychology and from its present entanglements with external systems of examination, the course of studies prescribed for the ordinary child would omit much that is now included and include much that is now ignored. Moreover, the order of presentation both of subjects and of the material in each subject will undergo extensive alteration. Changes in this respect have already well begun, and changes blindly initiated move towards enlightenment.

Things are moving towards a triple control of the individual life, one that roughly corresponds to the old control of the Church and State, with the added partnership of organized industry. To the modern representative of the

Church—whatever ultimate form this institution may take—belongs the function of guiding the individual for his own personal good. It is a service which the Church might easily have continued to perform, but for its insistence upon doctrinal terms. What was in effect a kind of strike of the religious ministry, met by boycott on the side of the general public, has led to the development of black leg service on the part of psychoanalysts and disinterested teachers. At present the organization manifests all the muddle of an emergency supply. Children in overcrowded schools cannot obtain adequate individual guidance, and the work of the psychoanalyst is curative only in exceptional cases, and not, as it should be, universally preventive. Nevertheless, feeling moves in the direction of agreement that something more must be done, though as yet nothing like a constructive policy of personal guidance has emerged.

The activities of the State in this connection are likely to increase. In earlier days a man could be born, could live and die in comparative privacy. Now, official notice is taken at least of his birth, his marriage, his income, most of his crimes, some at least of his diseases; and his death, too, is registered. The registration of particulars about the individual will extend to educational records, abilities and disabilities, and of all that is relevant to his social life. Much of this will be done by institutions of the kind which our present " vocational bureaus " dimly and crudely foreshadow. Ultimately a public service more or less State-controlled will be developed to mediate between Education and Industry. When life becomes more fully organized into a system of interlacing rhythms, the unit of the secondary Educational System will be an institution embracing factory, school, and clinic under a single board of control with its industrial, educational and medical representatives; and the threefold plan will no doubt be adopted in both the higher and lower grades.

Early in life some of the main tendencies and abilities of the child will be diagnosed and he will be sent to the appropriate school and industry.

Doubtful and obscure cases will, with the " allround " types, be sent for a " general education ", in the course of which, however, they will come under more careful examination. By the age of ten or twelve the child's future course should in outline and general character be fairly clearly known, and by sixteen he will have become more or less specialized to a certain type of course. Not, however, completely specialized. To meet the problem of trade-fluctuations and consequent unemployment, and as a preventive of industrial monotony, each person will probably be taught a variety of pursuits, achieving a balance appropriate to each case of mental and manual work. The system of multiple vocations would probably be operated in conjunction with some form of industrial conscription; for increased wisdom in government should by this time have undermined popular prejudice on this score.

At the prospect of such extensive changes in our educational institutions one wonders what is going to happen to our ancient and venerable universities. They might, consistently with all that has been said, preserve their functions as specialized seats of learning; but it is more likely that long before a rational system of education has been evolved they will have changed their character beyond all recognition. In optimistic, but perhaps short-sighted, moments, it is commonly supposed that with the increasing demand for education their future is assured, and that they will assume ever-increasing importance and enjoy ever-growing popularity and respect.

Let us, however, for a moment consider the instructive parallel of the Holy Catholic Church, the unfortunate history of which is a byword at the present day. The causes of its decline do not, I think, include one that is frequently suggested. Men are not to-day less religious or less in need of its holy offices. The change has come about not by decreased demands but by competition and an augmented supply. Men still need the things of which the Church once held an almost complete monopoly. They need some sort of philosophy, some personal guidance in the difficult art of the moral life, and many other things which once the Church alone could give.

The invention of the Press made possible the novelist. The novelist found himself adopted as guide, philosopher, and friend. The younger generation now-a-days acquires its general outlook from the novelists and playwrights, who exercise an almost sacerdotal influence. Personal guidance is offered by the psycho-analyst. He exercises the priestly functions of the confessional and of exorcizing demons; whilst Coué and his disciples offer a new technique of prayer. The validity of these practices is not here my subject, but the fact that people are getting what they want. Such new ways of meeting old demands are continually being found, and may not the Universities soon find themselves in the position of the Church? Several things would make the suggestion plausible.

In the first place, education is already being purveyed as a commercial product. Quite apart from the self-educators produced fortnightly by the Press, there are Mind-training Institutes and Correspondence Colleges. Universities at the present time can afford to ignore this competition. They have, they think, the pull in traditional prestige, the advantages of the tutorial system, the general atmosphere of culture, and they have, we trust, at present the most distinguished scholars. But traditional prestige is a slippery foundation for any institution now-a-days. The test that is being applied is that of practical efficiency. Already happy-go-lucky methods in University finance are being examined by Royal Commissions and by financial experts. The next thing will be that they will start to scrutinize the educational methods being employed. Already, moreover, the newer institutions are being forced to mould themselves upon the model of an up-to-date business office. Soon they will adopt those well-proven adjuncts of commercial progress — modern methods of advertisement, for instance.

When the printing press arose, priests, no doubt, preserved their self-assurance by reliance upon the peculiar efficiency of the spoken word. They clung to the sermon and scorned the weapon of the pamphleteers. When comparing themselves with the Correspondence Colleges, the Universities resort to a similar argument. One day, however, the Correspondence Colleges will realize the possibilities in wireless as an adjunct to their methods.

When with the assistance of its influential board of directors some national broadcasting company finally secures its charter, the following situation will arise. Phenomenal salaries will attract the most distinguished exponents of the various arts and sciences to the central broadcasting stations. District Colleges staffed by automatic invigilators will call the roll of students at the various lectures. These will be delivered by a loud speaker synchronized with a televisual cinema projecting the lecturer's blackboard and other illustrative material.

The economy effected by conducting classes of say half a million or equally easily ten million students will enable anyone to follow courses by all the most distinguished exponents of his subject. European students in psychology, for example, will attend say at 9 a.m. for Professor Stout, followed at 10 by Dr Sigmund Freud, and at 11 by Professor G. E. Moore. Nor need these distinguished men repeat their lectures until they have something new to say, for dictaphones will be included among the listeners-in, and another relay of undergraduates will acquire their wisdom from the broadcast-phonograph.

Such a soulless business organization, it will be objected, could never replace our present establishments of culture. But here, again, I feel the objection is of purely transitory force. Progressive differentiation which I have mentioned as an important aspect of evolution has produced the antithesis of commerce and cultured life. Differentiation, however, is to a great extent counterbalanced by processes of integration; and such integration we can see in progress in the business world.

The commercial applications of art besides influencing the dividends of our railway companies must be exercising a cultural effect upon their boards of directors ; and culture filters through in many ways. The justly renowned polish of the Oxford graduate is finding a lucrative field of expression in modern salesmanship. An up-to-date salesman now-a-days speaks considerably better English than his customer, and the average waiter is better mannered than the proprietor of the hôtel. It is inconceivable that the employer will continue to be quite uninfluenced by his cultured and refined subordinates. Perhaps this is the secret of the growth of those Learned Societies the purport of which we understand is to supply that liberal education demanded now by busy business men. When the captains of industry have mastered this business of being cultured and have seen its possibilities in practical affairs their own peculiar talents will be applied in ever wider spheres ; of which education will probably be the first.

V

Dr Schiller is of the opinion [1] that the evolutionary impetus is spent, and biologically at least we have seen the end of change. I wonder. And supposing he were right ? The process might begin again ; this time with the men of science at the helm.

The vast and powerful machinery developed to meet the needs of Scientific Management in industry, government, and education may make it possible to initiate and control the operation of the natural laws of change. Side by

[1] *Tantalus*.

side with the direct control of character, belief, and action, large-scale experiments in eugenics may proceed. Such experiments would not demand the clumsy expedient of the Act of Parliament. The triple control will provide adequate means for eliminating the intractable factors in the situation. Perhaps the constitution of the United States of America provides the most promising machinery at present for large-scale sociological experiment. These States, moreover, are fortunate in the character of their population. But whether by education or by eugenic control man himself will change. Physically, by a process of atrophy (at which Dr Schiller will not, I trust, demur), he may consist chiefly in a head from which will depend a mass of atrophied limbs. He will presumably be more or less permanently mounted upon a compact automobile, travelling by road or by air. No doubt a submarine attachment will be available to facilitate his morning bath.[1] But it is not with his appearance that we are here concerned. What is the future of his mind ?

There are interesting possibilities in the sphere of sense alone. Some of our present senses, such as taste or smell, may disappear altogether. More and more we seem to depend upon purely visual cues. On the other hand, vision and hearing may be

[1] But on all questions relating to our mechanical future I must refer the reader to Professor A. M. Low, to whose works *The Future* and *Wireless Possibilities* I myself am chiefly indebted in this respect.

considerably augmented. Even though they be supplemented by microphones and microscopes, we always strive to see and hear a little more than the best of instruments allows. If, after all, the biological vitalists are right, so long as we are interested in the constitution of matter there is a motive for further variation. But, as Professor Low points out, our æsthetic susceptibilities will have to be modified on the way. When we first begin to perceive the fauna in the cheese and the microbes on our walls, and when, like the birds, we go into the country to listen to the worms, we shall have to revise our values. Our poets, too, will find new themes to sing.

Memory, even more than sense, suggests many possibilities of development. By hypnotism, or by the natural working of the unconscious mind, it is said that we can recall even the earliest events of life. Soon, perhaps, this power will be under conscious control, and the autobiographers will have their millennium too. But, quite apart from this, new methods of education and a little Scientific Management would enable us, even now, to do much more with our memories through the powers we already possess.

The developments of intellectual life are the most difficult to foresee. But there are a few clues which enable us to guess at their possible nature. What purely intellectual achievements have been made in the past seem to depend upon the development of language and the elaboration of symbolism convenient to thought. Philosophers and scientists have continually expressed dissatisfaction at the modes of expression at their disposal in a language primarily devised to express our emotional and practical demands. Scientists have invented a technical terminology in their nouns. Mathematicians go somewhat further, and have invented more thoroughgoing systems of symbolism. Philosophers have always found it necessary to invent peculiar manners of speech, and now are seriously beginning to follow the mathematicians. It is possible that quite a new language will arise, probably a variety of purely technical languages—useful for different purposes of thought. The process of differentiation manifests itself here as in other ways. Already people following different pursuits find it almost impossible to converse, at least with anything approaching mutual understanding. And this process of specialization which education and increasing knowledge forces upon us points to another obstacle in the way of the Esperantists and other propagandists for a universal language. By the time this language is adopted, we shall have lost all common interests for the discussion of which it might have been employed. Perhaps our newer universities will inaugurate professorships in small-talk to keep alive an obsolescent art and to correct this serious menace of specialism.

More rapid than changes in the

intellectual life will be those of character and of temperament. There are two significant facts in the trend of man's emotional development, one of special and the other of general interest.

Firstly, there is the curious decay of cruelty and the increased desire to avoid inflicting pain. No one, surely, would deny an enormous change in this respect even in recent centuries. And we are beginning to discover the lengths to which development may go. Even now it often causes a moral man to attribute base motives to himself in doing a kindly act to spare his friend the subtle pain or sense of obligation which occurs in the feeling of gratitude. Modest forms of generosity often have this motive in their modesty.

There is not much further to go before we reach the social life of Henry James's novels—where the principal characters have exquisite consciences, where good souls are tortured by the regret that they had not expressed their reproaches a shade or so less directly so as more to convey the impression that they had really said nothing at all.

Now-a-days you are afraid to forgive your enemies, lest the pain of remorse you thrust upon them should exceed the discomfort inflicted by a direct retaliation. The finer spirits of the age cannot live up to their own ideals because of the sense of inferiority it would impose upon their less advanced companions. Society in the thirtieth century will assuredly be even more refined.

The second significant fact of the moral history of our age to some extent ameliorates the prospect. We are gradually growing out of our emotions. We are acquiring self-consciousness and becoming habitually introspective, and, as the psychologists tell us, you cannot observe an emotion without decreasing its intensity.

This partly explains what is curious in what is called "modern love", as practised and expounded by our younger novelists. It is essentially introspective. Every schoolboy knows what is going to happen when he falls in love. He knows about the illusions arising from emotion. The undergraduate cannot make love extravagantly like the Elizabethan poets, or sentimentally like the Victorians. When he feels the passion rising, he informs the provoking cause of it, that he knows quite well that, in spite of it, she is a perfectly ordinary little person with nothing very much to recommend her. They both agree that they are victims of an obsolescent mechanism designed for biological purposes, a mechanism primarily intended for the brutes, but annoyingly persistent in its control over civilized man. They agree, however, to act the performance through, because a rational way of making love is difficult to devise. They agree to make senseless and meaningless remarks. He is prepared to rave about her perfectly beautiful eyes whilst rationally convinced that they show a distinct suspicion of squint. She in her turn will admire his manly form whilst fully conscious that he is rather undersized and remembering that he signally failed to get his Blue. The kind of objective attitude that we are being taught to adopt towards criminals, which the psycho-analysts tell us to adopt towards the libertine, sooner or later we shall have to adopt towards ourselves.

In any attempt to picture to ourselves the life of future man one of the chief of the things which we have to take into account is the entirely remodelled constitution of society. There will be a small upper class—very small because, as all the prophets agree, the decrease of the birthrate in the upper strata will assume for sometime quite alarming proportions. It will be regenerated for a while by the bolder, more adventurous, and more capable of the working-class, who use their wits to escape from the situation into which they were born. Later, of course, this escape will be facilitated by reliable mental, moral, and vocational tests. The working-classes proper will then consist only of those less capable of administrative, organizing, and managemental functions. There will be relatively little unrest because the ruling classes will have learnt the wisdom of making the workers comfortable and happy; and potential agitators by the system of vocational selection will find themselves more lucratively employed in voicing the opinions of the ruling classes at the broadcasting stations. Plato will have come into his own.

His criticism of democracy will have become accepted and the philosopher kings—with a difference—will have become established in the seats of power.

Whilst the masses enjoy their innocent pleasures, the ruling caste will live a life of its own. Judged by the present, one might expect it to be a life of sport, gambling, and house-parties, and there may be a transition-period of this kind. In this no doubt some curious forms of sport will be devised. If the theory is well founded which asserts that our present modes of sport are debased survivals of ancient occupations, the craze of the fortieth century may be the fascinating game of mining. It will be played in specially constructed coal-mines, filled with ingenious traps, blind alleys, and water-hazards. Messrs Pope and Bradley in those days will circularize their clients by provocative letters pointing out that last year's mining suit is hopelessly out of date, or that the time has now come for replenishing the aerial wardrobe for the forthcoming polar sports. Chemical warfare, conducted with novel forms of laughing-gas, will be a popular institution at Christmas parties.

Various considerations combine, however, to suggest that sport is but a transitory phenomenon in human life.

Firstly, the applications of scientific method to its pursuit is tending to remove its distinctive character. And this tendency for economic reasons is likely to prove irresistible. The old-fashioned sportsman will protest against the newer methods. He may refuse to bring to the golf course wind gauge and slide rule ; he may refuse to attend to the lessons of motion study, but in consequence of his stubbornness he will find himself too ridiculously outclassed to appear on the course at all. His will be the fate of the older militarists who protest on similar grounds against chemical warfare.

To those who play the newer game, sport will become a rigorous discipline and a scientific pursuit.

The general public, moreover, will increasingly enjoy its sport by proxy. The televisual cinema, which will then be stereoscopic, will provide a substitute for the football match ; and the sporting spectacle will naturally form only an incident in the variety show. Again, the sporting interest will be merged, this time in the interests of general entertainment.

Another important change already far advanced which will radically alter the character of sporting life is man's recent discovery of his mind. The punch ball and gloves now hang limply on the wall whilst the one time early riser lies comfortably in bed cherishing the formula by which he grows sturdier and more Spartan in every way. Instead of Sandow we have Pelman, and this will spread to the schools. Enterprising teachers graphically plot the progress of their pupils upon the wall. Small boys will exhibit their I.Q.s and scorn a display of biceps. If Alec Waugh continues to write about it, the change may reach the Public Schools. Again, it is the work of the printing press which is at last beginning to tell. Anagrams and intellectual puzzles of all kinds have existed for the few from ages immemorial, but in the cross-word puzzle the principle is finding a wider appeal.

Lastly, the final blow to sport and to our British system of thought built up upon it will come from the industrial field. Industrial psychology will undermine it by eliminating its fundamental function. The forms of sport with which we are most familiar arise chiefly as a reaction against sedentary and ill-planned occupations. Scientifically organized labour of the future will exhibit more variety and give fuller expression to the many-sidedness of man. Excessive specialization of the individual in his work has produced an artificial differentiation in the arrangement of his life, of which the distinction of work and play is the most glaring example.

With the outgrowing of sport, or, more strictly, with the blending of work and play man will apply himself to tasks commensurate with his powers. Dimly we begin to see what these tasks will be. The task of government, complex as it may be, will be less difficult in many respects than it is to-day. The danger of revolution should be only a remote contingency. The stability of a society ultimately depends inversely upon the prevalence of nepotism. A government of the efficient and the talented equipped with scientific knowledge would be impregnable. It matters not that men are mostly irrational so long as the rational have the power to rule. The irrationality of the rest is subject to uniform law, which, understood, can be brought within control. But the temptation to the ruler to hand on the reins of power to the inefficient with whom he has emotional ties of paternity or friendship is almost irresistible. The difficulty, however, is not insuperable. Some form of "Socialism", which started all at equality and promoted to power according to ability, gauged quasi-mechanically by mental and vocational tests, would be a partial safeguard. Moreover, in a compact ruling class a kind of family moral tradition might emerge. This would be reinforced by the commonsense recognition of the dangers of promoting the inefficient on a large scale over the heads of the competent. The obvious effect of such promotion is, in the long run, revolution; and revolutions will probably be recurrent until the lesson has been learnt.

VI

And will these god-like beings with their infinite wealth and power, in their perfect rationality, be any happier or even morally better creatures than ourselves?

The formula for happiness is a fairly simple one. Its relative components form a vulgar fraction. The denominator is constituted by that towards which we aspire; the numerator is the number of our achievements. There is a law by which these two vary almost in unison. Any augmentation of achievement sends up proportionally the number of our desires, whilst great losses are followed by resignation. It is just for those fleeting moments in which the balance is disturbed that we experience bliss or misery. But the moments are very fleeting, and the normal ratio is restored. Happiness is the carrot before the donkey's nose. Therein lies the moral of Tantalus, so grievously misinterpreted by Dr Schiller. Tantalus is the symbol of human progress. The only danger is that under Scientific Management man may see through the trick.

And will he be better then than he is to-day ? That, however, is a question for the moralists of the future to decide.

ARCHIMEDES

OR

THE FUTURE OF PHYSICS

BY

L. L. WHYTE

LONDON:
KEGAN PAUL, TRENCH, TRUBNER & CO., LTD.
NEW YORK: E. P. DUTTON & CO.

To

LOTTE

Made and Printed in Great Britain by
M. F. Robinson & Co. Ltd. at the Library Press, Lowestoft

CONTENTS

ARCHIMEDES

OR

THE FUTURE OF PHYSICS

CHAPTER I

The Sciences Converge

One of the most fascinating features in the history of thought is that on several occasions an important new idea has come simultaneously to independent minds. Thus after Euclid's geometry had remained without a rival for two thousand years the conception of an alternative non-Euclidean system was reached separately by Gauss, Lobatschewsky, and Bolyai during the years 1820-30. Bolyai's father, while ignorant of the fact that Gauss had already made the same discoveries, wrote to his son urging him to publish his results and used the following prophetic words:

"There is some truth in this, that many things have an epoch, in which they are found at the same time in several places, just as the violets appear on every side in the spring."

Another example of the simultaneous emergence of an idea in the minds of different thinkers is given by Darwin in his introduction to the *Origin of Species.* He there calls attention to the fact that in 1794-5 the broad idea of the evolution of species—though not its cause—was simultaneously formulated by Goethe in Germany, St Hilaire in France, and his own grandfather, Dr Darwin, in England. Moreover Darwin himself had the remarkable experience of finding in an essay submitted to him in 1858 by A. R. Wallace a complete summary of his own unpublished theory of natural selection as the chief cause of the evolution of species.

The last few years constitute another critical period of a similar kind, since an idea, which when made precise will transform scientific thought, has already come independently to many thinkers. Since 1922 many scientists have felt that in studying the emission and absorption of light physics has come near to the problem of life.[1] Others have proposed that in order to straighten out its atomic problems physics will have to take a hint from biology, but what this hint should be has not yet been indicated. The following pages suggest a definite line of advance for physics, and interpret these isolated flashes of intuition as evidence of a special feature in the present situation of the sciences.

We stand at the eve of a new epoch. Physics, biology, and psychology are converging towards a scientific synthesis of unprecedented importance, whose influence on thought and social custom will be so profound that it will mark a stage in human evolution. For centuries science has concentrated its highest genius on the study of inanimate matter; to-day the three great sciences are at last reaching the problem of life. For their researches on matter, life, and mind are now overlapping at one common issue: the nature of the fundamental electrical processes which underlie radiation and chemical combination.

Thus physics is at present occupied with the changes that occur when an atom emits either light or electricity. Biology is at the same problem in studying the electrical processes which are the basis of all organic behaviour, whether in primitive forms of protoplasm or in the highly developed central nervous system of man. Meantime psychology is dealing with an identical process when it analyses the structure of mind, and considers the elementary changes of consciousness which are produced when light of a given colour falls on the retina and sends its influence to the brain.

As the result of these convergent researches, life and consciousness will soon be subject to the first stages of a theoretically-grounded control, compared with which the present tentative efforts of medicine and psychology will be looked back on much as we remember the haphazard work of the alchemists before the foundation of chemistry. But this development of human knowledge and powers will carry with it great responsibilities, and scientists have to prepare themselves for the new tasks that will very soon fall to them. By indicating the main ideas through which this broad scientific synthesis may come about, this essay aims at showing that this possibility has to be taken seriously. We shall first examine the situation in physics and then turn to consider the influence which future developments of physical theory may have on biology and psychology.

Two main types of process defy interpretation within the present scheme of physical conceptions: life itself, and the atomic processes of radiation and the building up of stable compounds. In organic processes on the one hand, and the energy-interchanges of atoms on the other hand, we find something happening which cannot adequately be explained as a change in the *structure* of the system considered. By structure is meant a spatial pattern of particles, which are supposed to be permanent and to move about like cricket balls or planets. Systems with a structure of this kind could not display the purposive quality of organic behaviour, and when we try to make a structural model of the atom we find that it fails to explain why the atom radiates energy in the abrupt packets which are called 'quanta', instead of in a continuous wave. We shall return presently to the question of organisms, after making an endeavour to discover why the atom cannot be described in terms of a particle structure.

In 1911 Rutherford achieved remarkable success in accounting for the results of his own researches in radioactivity by adopting a model of the atom as a miniature solar system, with planetary electrons rotating rapidly around a nucleus. But in order to explain the fact that the spectrum of the light emitted by an atom shows a characteristic series of lines, Bohr suggested that an electron inside an atom could emit light only by making a discontinuous jump from one possible orbit to another quite distinct orbit. This apparent discontinuity in the motion of electrons has intrigued physicists for more than ten years, and the following interpretations have recently been offered for this puzzling behaviour:

1. Nature is made up of electrons,

but neither space nor time is fundamentally discontinuous. The electron appears to have some freedom of choice, and to be able to reappear unexpectedly at forbidden places.

2. Nature is not discontinuous or arbitrary, but nevertheless something prevents us determining all the things we should like to know about an electron. For instance, if we try to determine exactly where it is, it behaves so that we cannot simultaneously measure its exact velocity. (Heisenberg.) This view may perhaps be interpreted to mean that we have made the atom model more complex than the atom itself is, and that consequently we have been using more quantities than are necessary for describing all we can observe of its behaviour.

3. Nature is not made up of electrons, but of waves. The atom must be considered as a system of electric waves spread over its whole volume. 'Electrons' are merely an inaccurate way of describing some of the properties of these waves. The wave picture of the atom

is, however, to be considered only as a temporary expedient to be used until some better description of the atom can be invented, in which both the wave and the corpuscular properties of atoms will appear as aspects of some more profound physical property. (Schrödinger.)

The first alternative is a mere cry of despair, since it does not propose any line of advance. But the other two suggestions may be combined thus:

4. The view of the atom as a structure of Newtonian particles is wrong since it gives rise to discontinuities, and provides more quantities than we at present need. A new formulation of atomic processes must be found using fewer quantities which will explain why we find wave properties, and why sometimes the electron does behave like a small billiard ball though really it is some different sort of thing.

Now since the Newtonian mathematics of moving particles is inadequate for describing the changes that go on in the atom—just as it is for describing organic processes—there must be some assumption implicit in Newton's laws which is valid neither for atom nor for organism. Such an assumption can be found very easily, though physics has never given it much attention. It is that the elementary processes in nature are *reversible,* or would be if they could be isolated. By reversible is here meant that the laws governing the process remain unchanged when the direction of time is reversed, i.e. when $-t$ is substituted for $+t$. If the law is changed by this substitution so that the reversed process never occurs or is recognizably different, then the process is called irreversible. An irreversible process can therefore be used to yield an objective criterion of past and future, when these terms have been once defined.

To take an example. If I am standing behind a hedge and take a cinematograph film of a stone which suddenly rises in the air and disappears from sight, I could not tell from an examination of the film which way to wind it. Thus if it is wound one way the stone appears to rise, and if wound the other way to fall from the sky. To tell which was the right way I should have to use my subjective sense of the direction of time, i.e. remember the fact that I saw the stone low in the air before I saw it high up. This case, like every gravitational process, is reversible, and motions of this kind have provided the basis for modern physical conceptions.

But suppose that instead I had taken a film of a cup of tea as it was cooling. One end of the film would show the steam above the cup and the spoon changing in length as it changed in temperature. Passing along the film these effects would grow less marked until the successive photos showed no variation when the temperature of the tea was nearly that of the surrounding air. It would be obvious which way to wind this film, without using any subjective criterion supplied from memory of the individual process which had been photographed. This process is irreversible, but physics has hitherto assumed that all such processes are merely the statistical result of a chaos of molecular motions each of them perfectly reversible.

The assumption of reversibility seems to some physicists so fundamental that they think there could be no science without it. But that is a mere prejudice arising from the fact that Newton conceived one particular way of giving mathematical formulation to the measurable features of physical processes. By suggesting that all the laws of nature might take a form similar to his law of gravitation, he made the implicit assumption that all elementary processes were reversible. Gravitational motions are so, at any rate within the accuracy of Newton's law, and as a consequence of the confirmation of his law and the fact that it has been taken as a model for the whole system of modern physical conceptions, the latter are only appropriate for reversible processes.

Apparent irreversibility, such as the cooling of a cup of tea, is attributed to statistical effects, and the second law of thermodynamics, which asserts that temperatures tend to uniformity, is treated as merely a statement of what is highly probable. This is probably quite legitimate, but even where no statistical effect can enter and the process is clearly irreversible physics usually adopts any measure rather than assume that a fundamental elementary process is irreversible.[2] We cannot be surprised at this, since if physics once admitted that any elementary process was irreversible it would have to give up the whole system of Newtonian conceptions. Matter, force, energy, action, and wave properties are all unsuitable for the treatment of irreversible effects since they all ultimately depend on Newton's reversible law.

An entirely new set of ideas is necessary for describing processes which necessarily proceed in one direction, so that one particular state of the system must precede another state. It appears conceivable that an alternative set of conceptions to replace the Newtonian might be established by demanding the irreversibility of all natural laws, as well as the demands hitherto made by physics, i.e. the permanence of matter and the conservation of energy.

The question of the reversibility of natural processes provides the key to a great intellectual struggle which is now in progress behind the complexities of philosophic and scientific thought. The issue can be formulated thus :

Is there a real temporal process in nature ? Is the passage of irreversible time a necessary element in any view of the structure of nature ? Or, alternatively, is the subjective experience of time a mere illusion in the mind which cannot be given objective expression? These are not metaphysical questions that can still be neglected by science with impunity. For just as Einstein made his advance by analysing conceptions such as simultaneity, which had been thought to be adequately understood for the purposes of empirical science, so the next development of physical theory will probably be made by carrying on the analysis of time from the point at which Einstein left it. Moreover, the above questions may be put into precise scientific form by asking if the causal relations which are studied by science are symmetrical and reversible so that we cannot obtain from them any criterion by which to distinguish past and future. If, on the other hand, they are asymmetrical and irreversible, the laws of nature lead us on necessarily from what went before to what comes afterwards.

CHAPTER II

A modern duel: Einstein and Eddington v. Bergson and Whitehead

In this battle over the importance of time and process great names stand out as representatives of the two opposed views: Einstein and Bergson, with their lieutenants, Eddington and Whitehead. The two leaders use very different methods. Einstein, as mathematical physicist, suggests that physical laws can best be expressed if we assume that space and time are so similar that physics can make no absolute distinction between them. Thus in relativity theory the symmetry of space involves the symmetry of time, and therefore the reversibility of physical laws, as has been shown by Birkhoff. Bergson, as biologist and philosopher, denies that the view of time which is implicit in relativity mathematics is adequate when a wider range of experience is taken into account.

Einstein starts by excluding all but a very narrow range of physical experience, and finds that he can make successful predictions about light and gravitation by treating the irreversibility of the passage of time as of no importance for scientific measurements. Bergson, by studying a wide range of biological and subjective experience, comes to assert the existence of a creative process, though the inherent limitations of the intellect and of science may leave the essence of this process outside their reach.

Both protagonists have left their flanks exposed, by omitting to present their view as a consistent logical system, Einstein because he is concerned only with the equations that can be empirically tested, and Bergson because his chief interest is non-intellectual. It is here that their lieutenants step forward to develop the two points of view, and hence to intensify the conflict.

Eddington provides a logical basis for the theory of relativity and reveals that the significance of physical laws is not quite what we used to think. They are, he argues, identities which the human mind discovers in its search for something permanent that it can call *matter* beneath all the changing appearances of the world. We have made matter the real thing by demanding permanence or indestructibility as the basis of physical reality. Now that we know that we have done this it need not trouble us too much to find that absolute unchanging matter doesn't exist, since this merely means that we started out with a demand that nature cannot fulfil. Unfortunately Eddington doesn't discuss what alternative demand we might now make in order to build up a more satisfactory system of scientific ideas. But in spite of his enthusiastic support of Einstein's theory, with its implicit assumption of reversibility, Eddington hesitates at least once in his advocacy of reversible laws, for facts are turning up which suggest that this undiscussed presupposition may not prove valid.[3]

Meantime Whitehead has been at work on the other side, and by sharpening his logic till few can understand him has made the idea of temporal process the basis of all intellectual and scientific thought, whereas up to now process has always presented many difficult problems for the intellect. He proposes that since the conception of matter has been found to be unsatisfactory we must start from the basic idea of process in building up a new physical theory. As a consequence of his line of thought, Whitehead found it necessary to reject some of Einstein's arguments and to show that Einstein's law could be reached from quite different postulates. For instance, Whitehead assumed that the motion of light was irreversible, and that light did not travel with the same velocity in the two opposed directions.

So much for one aspect of the conflict, its logical and philosophical basis. But the issue must be decided by appeal to experimental confirmation over the widest range of phenomena. Orthodox physics still assumes reversibility, and has on its side the explicit statement made by Einstein in 1925,[4] but by doing so it excludes at the start any reference to organic processes. Conceptions based on this assumption could never be legitimately applied to life, and all attempts made hitherto to explain the central controlling processes of organisms in terms of classical physics have necessarily failed. We know now that this failure could have been foreseen.

The same objection cannot be made

against the basic ideas of Bergson and Whitehead, nor against the new atomic physics as interpreted by Born, as we shall see in a moment. To Bergson and Whitehead, as to many others amongst whom Lloyd Morgan must be mentioned, the process of nature is creative, i.e. it involves the coming into being of the new, the appearance of new combinations essentially precluded before. This probably means that the laws of physics which are to describe what is actually happening in the world must be given irreversible form. For reversible equations make no distinction between to-day and to-morrow, and cannot express the fact that at later moments new forms may emerge, either in the evolution of organisms or of stars. On the other hand irreversible laws can be arranged so as to display time as an active factor in causation, i.e. to emphasize the fact that a certain period of time necessarily has to pass before some new combination can be attained.[5]

The upholders of a real process in nature can appeal to the facts of organic life, human memory, and to biological and stellar evolution. But their case is still weak because fundamental irreversibility has not yet received explicit mathematical formulation suitable for experimental test. When this has been done the intellectual battle will be brought to its decision, and if irreversibility wins the day biology and psychology will find themselves in possession of a physical basis well suited to the facts with which they have to deal.

There is reason to believe that the decision will be made very soon. We saw that the implicit assumption of reversibility underlies all Newtonian conceptions. It may therefore be that the reason why we cannot interpret atomic behaviour in terms of particle motions is that electrical and radiational processes are essentially irreversible. Particle motion and wave propagation—the two ideas on which all modern theories of matter are based—are both represented by mathematical expressions which are essentially reversible since time enters only through the square of 'dt'. If the quantum processes should prove to be irreversible, we have already found a reason why the old conceptions of particles and waves must be inadequate.

This speculation may indeed be found correct, since Born, one of the leading experts in Quantum Dynamics, asserts that all quantum processes are irreversible and that the apparent reversibility of classical processes is only an approximation due to the fact that their irreversibility happens to be negligible.[4] We may therefore hope that the atomic physicists will soon formulate the quantum laws in a clearly irreversible form which admits of precise experimental test.

But this may take some years, and in the meantime we must look around and see how this issue is affecting current thought. We find the doubt about process presented by Mr Sullivan (in *Gallio*), who has not yet made up his mind to which side science will grant the victory. Thus on one page he writes: "it seems to be true that events do not really take place, we come across them" and suggests that process may be "a totally irrelevant idea when applied to reality". But later we learn to our surprise that "it seems likely that (in scientific theory) the world will have to be regarded as an evolutionary process, where patterns of value emerge". However, this inconsistency need not bother us, since we are told that "the teachings of science so far as the spiritual problems of man are concerned are merely irrelevant".

These views reflect perfectly the uncertainty of the time, and will be looked back on as a precious record of the state of mind which preceded the scientific synthesis. Perhaps the most interesting feature of the essay is the indecision it displays with regard to the spiritual importance of science. This is a relic from the days when there were two worlds, the world of science and the world of religion and art. No one ever knew which of these worlds they were living in, and this is no wonder. For the division was made only because at one time it looked as though the scientific method could only deal with *quantities*, and therefore that science could have nothing to say about values or qualities. This view is no longer tenable. For instance, there is a quality in organic integration which most of us value, and without this and many other such conceptions biology and psychology could not get far.

Before proceeding any further it is necessary to correct a common misunderstanding with regard to the significance of Einstein's theory of relativity. This theory is mathematical, and is based on a series of postulates which rule out any claim to present an ultimate theory of space and time. One of these postulates[6] asserts that all our physical knowledge can be reduced to the space-time coincidences of pairs of point-events, or in other words the intersection of the world-lines of electrons. No respect for the supreme genius who predicted two experimental results and eliminated the chief discrepancies remaining in Newtonian theory should restrain scientists from pointing out that this postulate assumes something that has never been known to occur, and has no valuable reference to the world of physical experiment. The confirmation of Einstein's final equations cannot give any validity to this postulate. For it is difficult to think of any physical experience considered by theoretical physics which does not involve the perception of light or colour, and one cannot assume that the perception of light is a perception of coincidences.

Light varies in colour and intensity; coincidence in space is too abstract to account for an effect which is subject to variation. Moreover all physical experience requires a certain amount of time, and this fact is neglected if perception is reduced to the recognition of instantaneous coincidences. Even if these two criticisms are left on one side we still have to notice that Einstein's postulate rules out from the range of physics the important fact that many processes are irreversible. For instance, if we accept Einstein's definition of physical experience, then the interesting fact that radioactivity is only observed in the form of disintegration, and not also as the reverse process of a spontaneous building up of heavier elements from lighter, has to be left over by physics to be dealt with by some other science.

It almost always happens that the formulations of genius are exaggerated and form the basis of a pernicious orthodoxy, and it has certainly happened to relativity theory. Against a tide of exaggerated praise Whitehead, Larmor, and Bridgman, as well as some Continental astronomers, have debated the general assumption that the theory of relativity is adequate to its task, but those in whose hands the power of orthodoxy lies have not yet answered their criticisms in print. Neglect has always been the weapon by which orthodoxy has unknowingly hindered the advance of new ideas. But while this neglect is easy to understand, it is really remarkable that the postulates of relativity theory were not subjected to closer examination before it was made the basis of wide philosophical speculation. The experimental confirmation of Einstein's law of gravitation does not guarantee his postulates, since Whitehead has reached a similar law (identical within the accuracy of the observations) from different assumptions.

Einstein's profound creative intuition

and use of a difficult technique compel our deepest respect, but his work should never have been regarded as a *general* theory of time and space. Not only does he neglect the question of irreversibility but it is very doubtful if periodic processes can be made to fit into his scheme, as has been pointed out by Russell and Bridgman during the last year. Probably Einstein himself has never regarded his theory as more than a stage in the attempt to create a still wider physical synthesis, and we must not interpret in a broad sense his statement that one of the demands of his theory " takes away from space and time the last remnant of physical objectivity ".[6] This could only be true if physical time shared the absolute symmetry of space, i.e. if physical processes were all reversible. But there are processes from which we can obtain an objective criterion of the direction of time, and hence time does retain an element of physical objectivity as distinct from the absolute symmetry of space. One of the most interesting features in the future of physics will be the explanation of the fact that Einstein reached a correct law from postulates of limited validity, and in this connection Whitehead's alternative derivation may prove to be of importance.

CHAPTER III

Time in Astronomy and Physics

The real discrepancy between the world of physics and that of life lies in the fact that physics has never recognized the irreversibility of time, while this is fundamental to life. We may even feel a doubt if the ' t ' of physics has the same significance as the time of biology, evolution, history, and human experience. The physical conception of time arose from the practical utility of clocks for describing natural processes, and finally took the form of defining astronomical time in terms of the rotation of the earth. The day was in fact taken as an absolute measure of time, and this remained quite satisfactory so long as the laws of physics were found to take a simple form with reference to the time so defined.

But then a complication arose. The study of the moon's motion suggested to astronomers that the earth's rotation was slowing down, i.e. to account for the apparent motion of the moon they had to assume that the day was increasing in length. The theory of the tides revealed a possible cause for this slowing down in the tidal friction on the bottom of shallow water basins, for instance the rush of the Atlantic tides into the Irish Sea provides an appreciable frictional force retarding the spin of the earth. In addition to this slowing down there appears to be a very slow periodic variation in the length of the day such as would be accounted for by a rhythmic expansion and contraction of the earth's crust. The astronomers declare that our old measure of time is not only getting slower and slower, it is even varying rhythmically! It is clear that they have thrown over the earth as their definition of equal time intervals and have surreptitiously substituted something else. Yet one cannot discover any formal announcement of this, or find out if they realize that by doing this they have altered the theoretical significance of all physical measurements. In earlier days physics defined time in terms of a selected clock, and then set about finding the laws of nature. But the old ways aren't good enough for the modern astronomer who gives us our time and sets the clocks of our physical laboratories. He has reasons for disapproving of the earth, and has almost reversed the procedure. In order to save the laws of inertia and gravitation in connection with the moon's motion —and to a lesser degree in the cases of the planets and the sun—he has made these laws his standard of equal time intervals in place of the earth's rotation.

It is a curious situation, especially in view of the fact that Einstein's law, which has superseded Newton's, is not very suitable for use as an astronomical clock, as has been pointed out by Larmor. Perhaps the physicist will soon be able to use the atom as the theoretical clock for physics, and we can go on using the corrected rotation of the earth as our practical standard. There is a faint chance, for instance, that if physics can invent some way of measuring the minute time intervals along the track of an electron, then electrons might be used as giving the fundamental measure of time. Thus if the velocity of an electron were first measured by some indirect method the electron itself might then be used as a clock. But in the meantime the astronomers should make a formal announcement to the Royal Society of what they have been up to. It then might be found necessary to appoint a commission to discover exactly what physics is now doing. For by using an astronomical clock of the new type it is assuming classical laws while research-ing on processes which are already known to undermine the absolute validity of these laws. Theoretical physics cannot hope to clear up its fundamental problems until it has considered exactly what is involved in this suspicious procedure.

Like most professions, physics includes a good deal of bluff, but unlike the others physics is now occupied on a campaign to get rid of all pretence. For instance, physical text-books have been filled for twenty years with phrases of this kind: "an electron with a velocity of so many cms per sec." Yet the professors omitted to tell their students the awful secret that this hypothesis of electron velocities is one that has never yet received direct experimental confirmation. To-day a reaction has set in and the demand is being made that physical theory shall not make use of conceptions that do not correspond to directly observed quantities. Thus the latest theories of the atom have eliminated the idea of electron orbits because it was realized that these were nothing more than a mathematical trick for calculating something quite different: the wave-length of the light an atom can emit. In place of the orbits it is hoped to substitute something which only makes use of the directly-observed features of the atom, but this new picture is not complete.

Yet physics still makes use of ideas that have not been adequately justified. For though the idea of moving electrons has been removed from the latest atomic model, no substitute for it has yet been proposed for the case of electrons outside the atom. It therefore becomes very important for the experimental physicist to discover whether he can measure the distance travelled by an electron in a measured fraction of a second. As yet we have no proof that nature has not confused us by making electrons behave rather like moving particles, though really they are something different. In fact we have not yet made enough direct experiments to know even whether the dimensional system which is used for electrons is correct. Since no electron velocity has ever been directly measured we cannot be sure that the dimensions of the new constant ' h '—called Planck's constant—are really what we suppose, energy multiplied by time. Until a way has been invented of making a direct measurement of some *time* involved in electronic motions, it is impossible for physical theory to know how it should deal with the quantum processes.

When we realize how uncertain are the conceptions on which the whole of electron theory is based, we may wonder what is really known about the atom itself. Yet it is possible that we know more about the atom than we think, and that what are talked about as facts concerning electrons and radiation may really be better viewed as information about individual atoms and the way in which they influence one another. The emission of light is an atomic process, and we only know about light when it has reached some atom and been at least partially absorbed. Some un-understood change of condition occurs in an atom when it radiates and passes this changed condition on to another atom. The absorbed energy may cause chemical change, as on a photographic plate. But if a human mind is to become aware of this change of condition, then sooner or later, directly or indirectly, its influence must be passed on to an atom in the retina. We know very little about this change of atomic condition, and though it is usually called a change of the internal electrical energy of the atom this supposes more than we really know until some electron velocity has been directly measured. The dimensions of electrical energy are taken as those of kinetic energy, i.e. mass times square of velocity, but we do not yet know if this describes atomic changes correctly. Since no one has ever measured a *time* involved in an electronic process, the scale of time in the atom might be quite different from that given by our calculations.

Our ignorance of what this change of atomic condition really signifies is so profound that some writers have begun to treat the atom as though it were an organism, alive when the atom is excited, and dead when in a state of minimum energy. Thus Whitehead proposes that we should call the atom an organism, though this of course may only muddle us since we know even less about life than we do about the atom.

Yet we do know one very interesting thing about this change which happens to atoms but cannot be reduced to a change of structure. When light reaches an atom in the retina, an electrical stimulus passes up a nerve and alters the condition of the protoplasm somewhere in the brain. This change in brain condition is known to us directly as the perception of colour. Therefore in one sense we know more about this change of atomic condition than we ever did about 'electric fields' or 'gravitational potential' or any other of the mathematical conveniences used by physics in correlating observed quantities. The change in a sodium atom when we put salt in a flame is not a change in the consciousness of the sodium atom, because it is not part of a complex nervous system with the same high co-ordination as is found in the human being, and therefore the atom has no consciousness. But when an atom in the brain undergoes the same change we may become conscious of it, and the changes in matter which occur when light is absorbed are undoubtedly associated with the problem of consciousness.

Thus we are led to ask: how are single atoms built up into complex systems which have the characteristics of life, and finally into still more complex systems which have human consciousness?

CHAPTER IV

An Evolutionary Experiment

Questions are often made unnecessarily difficult by their being expressed in an abstract or theoretical form, and instead of asking What is life ? it will be more valuable to put forward a practical issue for discussion : Could an infinitely wise physicist order the necessary chemicals to-day, and to-morrow put together a synthetic man ? If not, why not ? What are we really up against, that seems to put some aspects of life beyond our control ?

Let us watch this ambitious physicist as he enters his laboratory. He has started quite easily and has in a moment prepared some simple molecules from their elements. Now he has completed the first colloid that he will require, and is starting on his first organic synthesis. But his infinite wisdom does not give him eternity within a minute, and we notice that he is getting on more slowly. While the actual combination of the first molecules took only about a thousandth of a second, once he had the apparatus ready, the simplest colloid took about a second. The organic colloid has taken him about a minute; it seems that nature won't work faster than that. She has her own rhythm and won't be rushed. If we wait patiently till the end of the day our friend may have his first speck of protoplasm, and all the skill in the world would only have helped him to make more of it, not to have got any further in his game of evolution.

But look at him now! He is making a hasty calculation as though he had just realized some great secret of nature, and knew that he could never create his homunculus. We look over his shoulder and read :

Estimated minimum time required by the synthetic processes of nature to attain various evolutionary stages.

Starting from the elements, to	Minimum Time
Simple inorganic compound	1/1000 sec.
Simple colloid	1 sec.
Protein	1 hour
Primitive protoplasm	1 month
Simplest uni-cellular organism	10 years
Flagellate	1,000 years
Mammal, including *Homo sapiens*	1,000,000 years

This highly speculative estimate is based on suggestive facts. A certain amount of time is necessary for two atoms to approach one another and form a molecule. The time required will be greater if many atoms have to settle down together into some special arrangement. For instance, the metal silver is normally crystalline, but if silver vapour is condensed too quickly the atoms will not have time to arrange themselves, and it is found that they pile up anyhow into an amorphous mass.

Colloidal processes require even longer periods, because great clumsy molecules have to arrange themselves on the surface of the colloidal particles. In elementary forms of protoplasm the molecular patterns are still more complex, and yet more time must be necessary to get the molecules properly adjusted.

It is probable that only our ignorance prevents us from building up protoplasm, but that we shall require rapidly increasing amounts of time for each successive stage of evolution. This will certainly be the case when we have reached organisms which can only be rendered more complex by controlling their environment while they reproduce themselves for many generations. A higher organism cannot be built up directly; the molecular arrangements in its body can only be reached through the synthesis of some simple form of life which must then be allowed to evolve through countless generations. Organic heredity resides in molecular patterns which can only be built up by this very slow process of repeated reproduction. Thus it is *shortage of time* that our ambitious scientist is up against in his haste to create a homunculus. Only the synthetic alchemy of time can build up organisms, each bearing within itself a long heredity.

The estimates given for the minimum time required in each case are about a thousandth of the actual time taken in a laboratory experiment or in the history of evolution as known from geological records. It may have taken a million years or more for the first mobile cells to have developed from inorganic materials and a thousand million years for the mammals. Yet perhaps these processes might have gone on more quickly. The times given are mere suggestions of a minimum time which may be necessary under ideal conditions. We waste a lot of time adjusting the apparatus in a laboratory experiment, and in evolution there may have been stationary periods with little or no new development. But it seems likely that when we know more about it we shall discover that a certain time is required for the formation of organic systems of given complexity. In this sense we may say that then human spermatozoon and ovum carry within them the synthesis of at least a million years.

Only an International Institute of Evolutionary Research under the most stable of Leagues of Nations could hope to create an artificial man, and even then man could hardly take the credit, for Time would have done more than man. But with sufficient consistency of purpose man could do this, provided he learnt how to make use of every moment of the creative power of time, and never made a slip by which the accumulated treasure of the years (i.e. heredity) might be broken. How man would learn to value life, and how profoundly such an experiment might alter his view of human beings, each one a priceless miracle, fruit of a million years!

In twenty years' time scientific knowledge will be adequate for the beginning of this giant task, and we shall be subscribing our guineas for the foundation of the Institute. Time has created man; man may use time to create man once more. With a million years ahead of us before we reach the sensitive mammals, we need hardly fear criticism from the Society for the Prevention of Cruelty to Animals. We are simply going to allow life to evolve itself under ideal conditions with Switzerland as the State for Evolutionary Research.

It may happen that under such perfect conditions life will evolve more swiftly than it did on this rough-and-ready planet. But equally well we—or rather our descendants—may find that the Darwinian struggle for survival is essential for evolution, and then the nations would have to debate on the morals of reproducing the 'cruelty of nature' inside the World's Evolutionary Zoo. Perhaps a wrathful god will seek to punish mankind for attempting to build this ladder to the secret of life, this modern Tower of Babel, and amuse himself by watching the community of scientists stricken by a plague of inconsistency amongst their weights and measures.

The possibilities of such grand schemes have to be taken seriously. We are now highly self-conscious beings with a tremendous technique for research. Men with genuine creative imagination who reverence life must shoulder the responsibilities of the twentieth-century consciousness, and use scientific technique for creative not life-destroying purposes. One can imagine a growing fraction of the interest now given to war, other people's adultery, and greyhound racing, turned towards Switzerland, whence at critical moments wireless bulletins would announce that the first amoeba had just successfully taken nourishment. If we wish it, the future of science can be such as to recompense for its recent occupation with gunpowder. Governments would be powerless to make war if the physicists refused to make the guns and the Royal Society called upon scientists to go on strike until each war crisis had been settled by arbitration.

The problem of life may be seen in a new light if the speculations of the last section are accepted and we assume that a definite period of time is necessary for the building up of any living organism. For if this is so the laws which govern life must involve the age of the organism since some definite moment in its history. We might choose for this moment the instant when the parent spermatozoon entered the ovum in the case of a higher organism, or in the evolutionary experiment just described the age might be reckoned from the moment when the first elementary chemicals were combined into molecules. The point is that this whole evolutionary process must be described by laws which take into account the age of the system under consideration.

Let us take a very simple, indeed the simplest possible, example. If two hydrogen atoms having just the correct total energy for the formation of a hydrogen molecule have approached one another and combined, the law describing what has happened must indicate that at a definite moment the combination was complete and the process at an end. This is an example of an irreversible process, since the molecule does not *spontaneously* break up again. Moreover, the mathematical formulation of this process must include the definite age of the system at which the process was complete, this age being measured from some selected initial moment.

This process provides an interesting limitation to a principle put forward by Maxwell as the basis of physical science. He suggested that the laws of physics must be considered to be eternal and unchanging and that therefore they must be expressed in a form which does not contain the time explicitly. This means that for physical laws there can be no difference between to-day and to-morrow. The laws are concerned with small changes which systems undergo in small time intervals, and need not express any fundamental distinction between one moment and another.

Such laws cannot express the fact that anything sudden ever occurs which makes an essential change in the system as when two systems become one, or when one system breaks up into two. The laws of organic growth or the evolution of individual systems must display the fact that at a certain age of the system special things happen, such as the combination of two hydrogen atoms, or the attainment of maturity by an organism. Maxwell's principle puts a limitation on the form of physical laws which precisely eliminates the laws that would be appropriate for organisms. But there is no reason why a broader physics should not try to frame this new type of law that would be applicable to the history and development of individual systems, and it is probable that if this could be done the reversible laws of Newton, Maxwell, and Einstein would appear as approximations which were valid when nothing of special interest was happening, i.e. when only spatial movements were involved without synthesis, disintegration or the emission of light.

Laws of the Newtonian type which Maxwell had in mind assume that one can adequately describe the present state of a system without specifying its past history. But we cannot say anything very precise about the inside of a living organism, and it is found far more efficient to describe what is known of its past history. We do not try to say where atoms are in an organism; instead we mention its species, age, etc. Organisms might be defined as systems whose future behaviour is more easily estimated from their past history than from what can be known about their immediate internal structure. The most convenient formulation of organic laws will therefore be expressed in terms of the age of the organism and take account of how its life has been spent. These laws are necessarily irreversible, since the assimilation of oxygen or food is always going on in a manner which can never be reversed. Life is like a function which must always alter in one direction; when this development ceases life has disappeared.

The contrast of living and dead now

appears less important than the following classification of natural processes:

1. Processes which are reversible and whose laws can be expressed independently of the age of the system, e.g. gravitational and mechanical motions which do not involve light or heat.

2. Processes which are irreversible, the laws being best expressed in terms of the total time which has passed since some initial state, e.g. chemical combination, growth, evolution, radioactivity, and all changes involving light or heat.

Physics has always asserted that processes of the first type were fundamental in nature, and astronomy provided the ideal example in planetary motion. It was this assertion that gave rise to the essential issue behind the conflict of mechanism and vitalism. But if Born is right, and the fundamental atomic processes are irreversible, then the situation is completely altered. There is no longer a question of life being an arbitrary irruption in a world of mechanical law, since the laws of gravitation and mechanics must then be looked on as the limiting case, when the irreversibility is vanishingly small, of a whole series of irreversible processes which constitute the most important examples of the fundamental order in nature. This series would include the atomic processes connected with heat, light, and electricity, chemical combination, colloidal effects, organic growth and evolution, and the highly co-ordinated electrical processes which form the physiological basis of consciousness.

If this view is correct the atomic processes of radiation and chemical combination should be just what the biologist needs to build up organisms. Instead of a chaos of little particles obeying inverse square laws, the modern physicist offers to the biologist a new kind of atom with electrical and magnetic properties which cause it to build up stable compounds.

The biologist may reply: "Yes, but organisms have four chief characteristics, their behaviour is irreversible, and displays growth, memory, and purposiveness. If you tell me that your atoms obey irreversible laws, so much the better, because my organisms certainly do. But your crystals grow very differently from my cells and organisms, and you can't explain away the apparent purposiveness of all life."

To which the physicist may answer: "Suppose that two hydrogen atoms are some distance apart with the total energy necessary to make a molecule. If they begin to move towards one another under some attractive influence which they exert we display no surprise. But they are moving towards a final end, which is an end, even though they are of course unconscious of it; and provided that nothing interferes they will reach one another, form a molecule, and the process will be consummated. The atoms move under an irresistible law of attraction towards a final condition which is unavoidable unless outside influences prevent it. The system of the two atoms develops necessarily towards a consummation, and the process has in this sense a teleological quality, though this need not mean that any god or man had consciously planned the end for these particular hydrogen atoms.

"This quality was not present in Newton's law of gravitation precisely because it failed to say what happens at the end of any process, for instance when a meteorite hits the earth. Newtonian laws avoid the responsibility of dealing with all the exciting events, like the wedding of the atoms or the death of the meteorite. On the other hand it appears probable that all irreversible laws can be interpreted as leading either from or to some critical end condition. Thus all heat processes tend towards an approximate uniformity of temperature, and chemical reactions also move towards a final condition.

"Such systems as these display the rudiments of unconscious purpose. One must imagine these systems made much more complex so that it takes a long time and considerable nourishment before their unconscious purpose is fulfilled, whether this be the instinctive reproduction of their kind or any other biological function."

"Maybe. I like the unconscious purpose which you have revealed in irreversible physics, because I am troubled by colleagues who see conscious mind everywhere.

"But if I grant that your view of the atom, and hence of molecules and colloids, allows me two of the four features I find in life, i.e. irreversibility and unconscious purpose, you have still to deal with growth and organic memory."

"Yes. Growth and memory are things that physics has as yet little to say about. But we have at any rate reduced the problem of life to smaller proportions. It is no longer the question what is life? but, how do colloidal processes build themselves up into continuously-active, developing systems which can react to their surroundings so that some distant condition can ultimately be attained? This is a much less difficult question. Moreover, since the problem of radiation underlies all the chemical processes which are associated with the maintenance of life, we may expect considerable assistance when physics has cleared up this crucial problem."

CHAPTER V

Physics and Mind

If a psychologist who was not a behaviourist had been listening to this conversation he might break in :

" Does the physicist seriously propose that we should try to leave mind out of our picture of the human organism ? Even if we can eventually explain the unconscious purposes of the lower organisms as ends towards which they are driven by physical laws, yet man has the supreme distinction of a conscious mind, he can select his aim, and if he likes renounce it again for something else. You must therefore allow in your picture for the emergence of mind at some point during the course of evolution."

"Wait a moment," replies the physicist. "Your whole outlook towards consciousness betrays not only an anthropomorphic standpoint, but one limited to a single stage in man's development. There is no single condition adequately described by the word 'conscious'. There are in fact a great many different states of awareness which may grade into one another, or may form a series of distinct conditions. We do not know much about them yet, but their variety is most striking. There is the dim sentience as we awake from chloroform, the awareness of the dreaming state, the passive experiencing that accompanies any intensely rhythmic activity such as running. Again, quite different states are known in day-dreaming, intellectual concentration and the delicately-balanced semi-consciousness of creative thought.

"Consider especially the states of awareness associated with love, or with the supreme creative activities of the mind. Free-will, or the deliberate choice of a purpose, is completely lost in a whole-natured falling in love, as it is also in the artist's need to follow some dimly-conscious intuition of a task he must attempt. At these important occasions free-will disappears before a sense of inner organic necessity.

" These examples seem to me to make it clear that 'conscious purpose' is not in any sense the ultimate or highest criterion of human behaviour, and that free-will need not be taken necessarily to mean the power to over-ride any laws of nature. In my view 'free-will' is simply the apparent characteristic of organic behaviour when no complete integration of the personality has been achieved and the mind seems to be able to oscillate from one purpose to another. We really have to deal in human beings with a whole series of forms of behaviour of increasing complexity and integration : reflex and instinctive actions, deliberate activity, and finally the intuitive whole-natured creative functioning which leads to ends which could not have been intellectually foreseen. To each of these must correspond a certain type of awareness, and in my view, a brain process of a definite degree of complexity. By analogy with our own experience of different modes of consciousness, we may be able to infer from the structure of the central nervous system of an organism what sort of awareness it can experience.

"Eventually we must expect to be able to give a complete scheme of all organic behaviour in terms of the organic processes and their laws, but none the less it will remain a great deal more convenient in some cases to refer to what happens to human beings by using words that suggest their conscious experience. The behaviourist denies the scientific significance of all but the very barest elements of conscious experience, but of course he has to start from the human perception of light and colour. Science cannot get on without ideas which obtain their whole meaning from the qualities of conscious experience, and hence the extreme behaviourist position merely arises from a prejudice which prevents clear thinking. But as a campaign to put more stress on the direct observation of what really happens to living beings in terms of physical movements, behaviourism can only do good by bringing more unbiassed knowledge about life.

" My own interpretation of the question may be put in this way. The thing that is given in nature is a process in time. According to its complexity and degree of co-ordination an organic process has different degrees of awareness. There is no one condition called human consciousness, because the human organism can function with different degrees of co-ordination, and if we ask if an atom in absorbing light is conscious, the question has no definite meaning. But in a few years those who are studying the physiology of the central nervous system will be able to indicate how many steps of synthesis and integration occur between the simplest cell and the creative thinker, and to each of these stages will be ascribed a mode of awareness. But below a certain degree of organic complexity this 'awareness', will cease to be anything that can be consciously imagined by man, e.g. below the dimmest sentience one might allow an undifferentiated knowledge of mere continuance, based in turn on the rhythmic pulsation of the elementary cells."

"Your scheme is of course still rather vague, but in its main outlines it appears satisfactory", replies the psychologist. "But tell me outright, can mind influence matter? If I understand you rightly, you suggest that matter certainly influences mind."

"On the contrary, I do not! You are back at the meaningless questions on which philosophers have wasted much time. To ask if mind can influence matter does not mean anything until you know what you mean by mind and matter, and to a scientist that means knowing the laws they obey. Now, on the one hand, relativity and modern quantum theory indicate that there is no matter in the old sense of particles made of some unchanging stuff, and physical science recognizes atomic and other *processes* as fundamental in the place of 'matter'. On the other hand, you really mean by 'mind' one particular form of conscious activity: the deliberate selection of a purpose. Therefore to give your question real meaning I have to ask instead 'Does the conscious selection of a purpose alter the physical processes going on in the human organism?'

"But that is an absurd question. It is like asking: Does a dint in the outside of a hat *cause* an alteration in the shape of the inside of the hat? To which the only reply is that the dint on the outside is merely another way of describing the dint on the inside. There is no *causing* of the one by the other any more than if you fold a bit of paper you can say that the crease on one side causes the crease on the other side. They are identical and the double method of description used in the question creates a meaningless problem.

"'Conscious selection of a purpose' is one way of describing a particular process, and after this process has occurred the brain will be different from before. The old theories of the correlation or interaction of mind and matter presupposed that they were separate things in themselves. The important questions become quite different when one realizes that mind and matter do not exist independently, but that they are both somewhat inadequate ways of describing certain *aspects* of one organic process.

The spatial aspect of organic process is called the physical organism. The temporal aspect of organic process corresponds to the content of its consciousness. The physical body is a group of spatial characteristics. Consciousness is a system of temporal elements; memory, anticipation, deliberate repetition, creative longing, hope and fear are all things set in time.

" Professor Alexander has said ' Time is the mind of Space.' He attempts to explain space and time by an anthropomorphic analogy. It is a very suggestive idea, though for the searcher whose goal is the nature of consciousness itself it is more valuable to put it the other way round: mind is the temporal aspect of process, body the spatial aspect. But it is very important indeed to notice that we have not yet found the adequate terms for describing these two aspects of process. Matter is unsatisfactory for the spatial aspect, because there are no unchanging particles. But nor is mind sufficient for the temporal aspect, because there is a temporal aspect to the combination of hydrogen atoms and to chemical and colloidal processes, and yet we must not speak of these as having mind. When the new words for these two aspects are invented they will form the foundation of the scientific synthesis which I am expecting."

To which the psychologist may answer: "Well, at heart I have always been a thorough-going determinist like you, at least in dealing with my patients. Moreover I find it works, because I have always included in my picture of the patient a life-impulse of some sort, which can be influenced by my personality. Thus if the behaviour of my patient is absolutely determined, the conditions which determine what happens to him include some inner life tendency, and also the effects produced on him by all the people he meets.

" But if one attempts to formulate such an absolute determinism, or to apply it to oneself, one gets into deep waters, and I haven't the courage to try it. It seems you must be right at bottom, but that only a god could believe it without its upsetting his mental balance or his sense of moral responsibility."

" There I agree," replies the physicist, " as long as one does not simultaneously revise one's whole view of life in terms of this new organic knowledge. That is a very big task, but I should like one day to attempt it. Two things especially would attract me to such a revision of human values. One is that people who ought to know better still go about making moral judgments about their acquaintances. Now that we know how profound is the influence on a child of the treatment it receives during its first five years of life, moral judgments become rather old-fashioned and only show that the person making them has himself not yet learnt to find emotional fulfilment in healthier ways. An analysis of human behaviour along the lines of organic determinism might do something to show that moral condemnations, whether of bolshevism or of the sins of one's children, are never effective unless immediately accompanied by positive example or creative suggestion.

"But there is another more attractive reason why I should like to attempt this transvaluation of values. If organic determinism is valid, then the artist's aspiration to create is a natural consequence of some organic law. Creative aspiration may then be looked on as the natural destiny of certain human beings, though they no more know where they are going than did the two hydrogen atoms. But organic determinism allows us to understand why it is of no importance that the artist doesn't know what he is going to create before he does it. It seems that in some matters our organic body is wiser than ourselves, or rather wiser than our very immature consciousness. When we have developed our consciousness by the discovery of the organic laws of our own natures we may be able to make human life more beautiful."

CHAPTER VI

The Future of the Sciences

The preceding pages have very broadly indicated the way in which current physical researches may influence the scientific outlook on the problems of matter, life, and mind. The view has been put forward that we are on the eve of a profound scientific synthesis of which the main outlines are already determined. These general suggestions will now be made more precise in order to offer to anyone who is interested the opportunity of testing for himself some definite prophecies regarding the future of scientific thought. The forecast made here does not involve any supernatural reading of the future, but is based on tendencies already inherent in the different departments of science. For convenience it is expressed in the form of separate assertions concerning the future of physics, biology, and psychology.

1. Before 1940 a very remarkable simplification will be made in atomic theory, which will indicate that in quantum processes physics has 'touched bottom' and that—for the time being—we may consider that nature is not infinitely complex within the heart of the atom. The proof of this apparent if not absolute limit to the micro-structure of nature will take the form of the discovery of simple relationships between the fundamental constants of atomic structure, e, m, M, c, and h. (The electronic charge and mass, the mass of the hydrogen nucleus, the velocity of light, and Planck's constant.) Such relations are already known but are considered to be of no significance since they are ruled out by the accepted theory of electrical dimensions.

Yet this dimensional system is not based on direct observation, and the importance of these relationships will soon be recognized in consequence of experiments aimed at a direct determination of an ‘electron velocity’, in a curved track. ‘Electron velocity’ as calculated from deflection experiments will be found not to be the same as the directly measurable cms. per sec., and in the case of straight electron tracks, the measured velocity may be found to be always that of light, though this does not mean much since the velocity of light in one direction has never been measured.

As the result of the study of individual radiation tracks, for instance in the reflection of electrons by crystals, and particularly of any *time* measurements that can be made, a new system of physical conceptions will be built up appropriate to irreversible processes, which will be substituted for the Newtonian reversible system. The new scheme will probably be based on the conception of the atom, with its radiating electron tracks, as a natural clock which not only can be used to measure out equal time intervals, but also to yield an objective criterion of past and future. In order to make this idea, or at least one part of it, capable of empirical test the following hypothesis is put forward: The time-interval between any two point-events on any electron track is a simple function of the length and curvature of the part of the track between the two points. This hypothesis contradicts the current interpretation of electron theory on a point which has never yet been subjected to experimental test.

The conceptions which will be built up on electron velocity experiments will very quickly bring within one simple theory the facts of chemical combination and colloidal processes. For these depend upon irreversible effects connected with radiation and electrons, and will therefore be amenable to treatment by the new conceptions for the very reason which necessarily puts them beyond the scope of Newtonian laws.

2. As the result of the alteration in physical conceptions biology will soon cease to draw a definite line between inanimate and living systems. The normal characters of life will be recognized as appearing in steps as one passes up the series atom, molecule, colloid, protoplasm, cell, and through further stages to mammal and man. In each class of organism a central controlling process will be discovered and its laws formulated with some precision, in terms of irreversible electrochemical processes. The process which in each organism represents the co-ordinating factor and is the life of the organism considered as a unit may for instance be described in terms of a quantity which we shall call 'f'. 'f' would be such that so long as 'f' keeps on increasing the organism is alive, while if 'f' stands still the organism dies. The rate of increase

of ' f ' indicates the tempo or intensity of the organism's life. In a simple case ' f ' might be directly related to the intake of oxygen or food, and just as respiration and assimilation are irreversible, so is the change in ' f '. ' f ' must go on increasing, or else cease to represent any quantity in nature ; as soon as it ceases to increase the process to which it corresponds cannot be identified any longer.

The most important factors which influence the life-function ' f ' (i.e. which affect the central controlling process in any organism) will be known before about 1950, with the result that local rebellions such as cancer will not only be controllable, but easily prevented. Harmless methods for increasing the rate of change of ' f ', i.e. for increasing the *élan vital* of the organism, will be discovered, so that, for instance, the duration of child-birth will be reduced to a natural minimum. If child-birth sometimes takes very long nowadays, this is presumably because the woman's body is tired, exhausted, or partially poisoned by her mode of living, and by raising her vitality at the critical moment we may expect to be able to let the process go on at its natural speed. There must be some minimum time necessary for the act, since a vast number of complex organic processes have to complete themselves in a certain order, but probably this time is considerably shorter than that during which many women in this country have to suffer.

It is already known that the Mendelian *genes* which determine heredity are related to the rates of development of special processes in the organism, and a control over the life-tempo, or rate of increase of 'f', in any organism or group of cells within an organism, will provide a new method of tackling the practical problem of heredity. It is possible that hereditary tendencies to specific weakness or disease will be overcome by accelerating or retarding the rate of development of the human system at some special moment between conception and maturity.

Rejuvenation will soon be safe and efficient, but not as a means for attempting immortality. It will be socially recognized as healthy and legitimate only when undertaken to compensate for premature ageing due to specific repressions, illness, or anxiety.

The elimination of known diseases by a genuine science of life does not mean that other diseases will spring up perhaps worse than before. A theoretical science of life will know the meaning of all disease, and will not prevent one in such a way as to give rise to another. Instead of making campaigns against influenza or any other one disease, it will determine the conditions in which no disease can survive, and thus gradually eliminate all the organic diseases which attack the body.

But this does not mean the attainment of a hygienic Utopia in which human life necessarily fulfils itself. A balance will be made to the disappearance of cancer and syphilis, not by the arising of other diseases but as a result of the consequent increase in the sensitiveness of the human brain.

The supremely difficult task of the next hundred years will be to keep the mind of the race healthy and stable through a period of critical sensitiveness. We are in a transition stage of violent instability, of intense cruelty coupled with compassion (America), of blended love of liberty and need of discipline, of emotional religions and of wars—but we must hope that it will lead to some mode of life with greater inherent stability.

3. Psychology is now occupied with the discovery that the human response to perceptions is not additive, i.e. that the effect made by a group of sounds or colours depends on the pattern in space and time in which they are arranged. (*Gestalt-theorie.*) For instance, the effect made on a man by the individual notes

of ' God save the King ' when played in the wrong order is negligible, and bears no relation to his response when he hears the tune played in a cinema, and it reminds him of ' patriotism ' and the War. So far no scientific method has been found of describing when a group of elements is to be treated as a ' whole ' for the purposes of psychology, and this is where the greatest advances may be expected.

Most scientific conceptions have been based on the method of spatial analysis, i.e. the reduction, where possible, of a thing to its smallest spatial elements. Physics, biology, and psychology have all lacked the equipment to describe what makes the atom, organism, or the pattern function as a unit, and how we are to know if some group is a unit or not. The analytical method is fully developed, but we lack even the basis for a synthetic treatment. This leads some hard-headed scientists of the materialistic school who will ' stand no nonsense ' to assert that there is no such thing as 'synthesis', that this is a mystical idea left over from primitive anthropomorphism. Yet to any mind that is guided not by prejudice but by a simple search for truth, the fact of synthesis is obvious, though not yet properly formulated.

Here modern physics can supply a clue. Analysis is the method required in a search for instantaneous spatial structure; the synthetic method which we need must deal with the temporal history and behaviour of systems. The fact that the human being reacts in the ways he does to a tune as a whole is evidence of something in his history, that he has heard the tune often under certain emotional surroundings. The unity of any synthesis, whole, or organism is not an instantaneous fact explicable in terms of structure, for we can recognize this unity only from a continued observation over a period of time.

Physics can invent one law to describe

the approach of the two hydrogen atoms to form a molecule, and in doing so treats the two together as a unit. This suggests that the fact of organic unity is to be defined and formulated in terms of an irreversible law which governs the system as a whole. Thus a group of atoms, cells, or any other elements is to be called a unit when, and only when, one irreversible law can be found which expresses the behaviour of the different elements as contributing towards some common end, like the formation of the molecule in the case of the hydrogen atoms.

We can now draw a practical conclusion for the future of psychology, which is in great need of a moral principle to guide its treatment of disintegrated human personality. On the analogy of the two atoms, a human being is to be considered as a unity when his whole behaviour displays continuous co-ordination towards some end. But there is an important difference in the two cases: the atoms move towards an end which we know because it has already happened in history, whereas man's development is creative, that is it proceeds towards an end we cannot know exactly before it comes into being. Thus the parent or psychologist need not trouble if he cannot understand what his child or subject is aiming towards: so long as some consistency and harmony of functioning is apparent, the 'end' can be left to nature to look after, because such harmony *means* that the organism is tending towards some ultimate condition.

The psychologists of the future will therefore have to follow some principle such as this: their only legitimate aim is the maintenance and restoration of harmonious co-ordination of all the human functions, and no concern need be paid to ultimate intellectual or spiritual ideals. Of course if the person considered is apparently tending towards some degenerate condition, that is known to the onlooker because it is *not* new but a repetition of what many human beings have done before, then this tendency can be altered. At least, it can be altered if the onlooker can use his intuition to discover signs of repressed conflict which show that the immediate tendency is not whole-natured, but based on the repression of some more profound aspiration or desire. Then by bringing this repressed aspiration back into consciousness the degenerate tendency may be arrested. But this control over the lives of others can only be effectively exercized by the intuitive discovery that their present tendencies are not whole-natured.

Prophecy can never be scientific, and forecasting in the realm of science is perhaps the most dangerous form of intellectual acrobatics. Science must be thorough, and all vague speculation is its enemy. But there are moments when a profound revision is necessary, and amidst the responsibilities and rich appeal of daily life no one will undertake this task who does not believe that it offers an adequate reward to science and to man. To-day prophecy can call attention to unjustified limitations inherent in current scientific thought, and encourage the students of matter and of life to get together and try to discover the single system of natural law which we must believe covers both realms. It may even help them to find crucial experiments by which to guide their search.

The reward is certainly great. The indifference to the destruction of life which has marked recent years is no cause either for surprise or for despair after an epoch of orthodox and insincere religion coupled with an abstract science of matter. One thing only can guide humanity to a saner and richer life: the recognition and valuation of life. This can be assisted by science and art both revealing life in all its significant forms. But the roots of art have been destroyed by the domination of a science which had not recognized the significance of life within the realm of natural law. For great art can only arise from a profound reverence for life, whereas to the scientific mood of this period life appeared as an arbitrary impulse in continual conflict with the laws of matter.

Physics is now studying light. The radiant influence of light nourishes life and within human body forms the fabric of consciousness. We are alive and conscious, but our consciousness is immature for we do not yet know the laws that govern our own lives and thoughts. Yet it is certain that light, life, and consciousness are bound together by some undiscovered law. This secret of nature's alchemy is still hidden from us within our own bodies. By revealing it physics will create a new hope for man.

NOTES

1 Whitehead, *Science and the Modern World*. Eddington comes near to the same idea in an essay in *Science, Religion, and Reality*, 1925. See also Weyl, *Was ist Materie?* 1924, p. 84. It has also been expressed by others quite independently, though I do not know of other published references.

2 E.g. the irreversible motion of an electron in the field of a bar magnet is rendered formally reversible by the assumption that the magnetic field is due to moving electrons. Yet this assumption is highly artificial since it postulates electronic movements that have never been observed. In other cases irreversibility is eliminated by the choice of special co-ordinate systems. Some physicists now hold the view that irreversibility may be inherent in atomic as it is in organic processes.

3 *Internal Constitution of the Stars*, 1926, p. 56. Compare note on p. 44.

It may be convenient here to summarize the processes that give at any rate superficial evidence of their irreversibility: processes involving heat changes, or the radiation of light, or mass; the production of energy in a star, the motions of electrons in magnetic fields, certain types of atom-ion collision in mixed gases, processes dependent on retarded potentials, radioactivity, organic growth and evolution, and consciousness itself. Eddington deals only with the case of the emission and absorption of light, but suggests that the direction of time can only be deduced from statistical processes. This is the orthodox view, though it is very doubtful if it is valid now that the quantum processes are receiving formulation. In this connection, see note 4.

4 Einstein. Berlin Akad., *Sitzungsberichte*, 1925, p. 418. But Einstein's view must be revised in view of recent experimental results (e.g. Harnwell, *Phys. Rev.*, vol. 29, 1927. pp. 683 and 831), if these have been correctly interpreted. See Born, *Zeitschr. für Physik*, vol. 40, pp. 177-8; and Jordan, *Naturw.* 1927, p. 792.

5 The idea that time may be an active factor in causation has the mathematical significance that 't' (for the system in question) must appear explicitly in the formulation of the law, and not merely as the square of a time-differential found convenient for the correlation of a standard clock with a reversible process which is being observed. A law whose mathematical formulation involves 't' measured from some moment in the history of the system, gives an entirely new meaning to 't', though one consistent with the properties of the reversible Newtonian differential 'dt'. Such a law may claim to express the fact of historic, irreversible, duration, a feature in nature which is neglected by laws involving only 'dt' squared.

6 Einstein, *Annalen der Physik*, vol. 49, pp. 776-7, 1916.

For Product Safety Concerns and Information please contact our EU
representative GPSR@taylorandfrancis.com
Taylor & Francis Verlag GmbH, Kaufingerstraße 24, 80331 München, Germany

www.ingramcontent.com/pod-product-compliance
Ingram Content Group UK Ltd.
Pitfield, Milton Keynes, MK11 3LW, UK
UKHW021501240926
475702UK00035B/35
9780415463041